THE FIRST

N O N L I N E A R

SYSTEM OF DIFFERENTIAL AND INTEGRAL

C A L C U L U S

*

Michael Grossman

University of Lowell

1979

M A T H C O

Box 240 Rockport, Massachusetts 01966

ISBN 0977117006

"...we find in the history of ideas
mutations which do not seem to correspond to any obvious need, and at first
sight appear as mere playful whimsies
- such as Apollonius' work on conic
sections, or the non-Euclidean geometries, whose practical value became
apparent only later. "

Arthur Koestler
The Sleepwalkers
London: Hutchinson, 1959

First Printing, December 1979

P R E F A C E

The classical calculus developed three centuries ago by Newton, Leibniz, and their predecessors is a linear calculus in the sense that its operators are additive. The first nonlinear calculus, the exponential calculus, was created by Robert Katz and me in July of 1967. The operators of the exponential calculus are multiplicative.

Subsequently we discovered that the classical and exponential calculi are members of an infinite family of calculi, all of which can be described simultaneously within the framework of a general theory. We decided to use the adjective "non-Newtonian" to indicate any member of that family other than the classical calculus.

In 1971 we completed our Non-Newtonian Calculus[1], which is self-contained and includes a brief account of the exponential calculus, eight other specific non-Newtonian calculi, and the general theory.[2]

I decided to write a detailed account of the exponential calculus for three principal reasons: it is easy to understand, has many attractive features, and may prove to be one of the most important of all the non-Newtonian calculi.

As is the case with the classical and all the non-Newtonian calculi, the exponential calculus possesses the following:

a distinctive method of measuring changes
   in function arguments;
a distinctive method of measuring changes
   in function values;
four operators: a gradient (i.e., an average rate of change), a derivative, a
   natural average, and an integral;
a characteristic class of functions having
   a constant derivative;
a Basic Theorem involving the gradient,
   derivative, and natural average;
a Basic Problem whose solution motivates a
   simple definition of the integral in
   terms of the natural average;
and two Fundamental Theorems which reveal
   that the derivative and integral are
   'inversely' related in an appropriate
   sense.

When applied to specific functions, the operators of the exponential calculus yield numerical results that differ from those yielded by the classical operators. And, of

---

1. Non-Newtonian Calculus (Mathco, 1972).
2. In Non-Newtonian Calculus the exponential calculus is
     called the geometric calculus. My reason for changing
     the name is unimportant.

course, the exponential operators reflect different conceptions.

The classical derivative is constant for linear functions, but the exponential derivative is constant for exponential functions, which explains the name "exponential calculus." The classical derivative is invariant under all changes of origin in function arguments and values, whereas the exponential derivative is invariant under all changes of origin in the arguments and all changes of scale (or unit) in the values. Thus, the exponential derivative is scale-free with respect to function values, a feature that should appeal to scientists who, believing that 'Nature hath no scale,' seek ways to express laws in scale-free form. (Chapter 4 of Non-Newtonian Calculus contains a calculus whose derivative is scale-free with respect to arguments and values.)

As will become clear later, the arithmetic average is the 'natural' average in the classical calculus, but the 'natural' average in the exponential calculus is the well-known geometric average. (The quadratic and harmonic averages are the 'natural' averages in the quadratic and harmonic calculi, which are treated in Chapters 7 and 8 of Non-Newtonian Calculus.)

Although there are many excellent ways of presenting the principal ideas of the classical calculus, the novel presentation in Chapter 1 probably leads most naturally to the development of the exponential calculus in particular and to the non-Newtonian calculi in general. Included in Chapter 1 is a simple method of defining tangent lines without using limits or derivatives.

Chapter 2 contains a construction of the exponential calculus, which is introduced by relating how it was discovered as a result of a chance observation of a simple analogy between a well-known algebraic identity and the algebraic identity that plays a fundamental role in the classical calculus. In fact, each non-Newtonian calculus is based on a simple algebraic identity, whose extension to continuous functions is called the Basic Theorem.

Chapter 3 includes the exponential arithmetic, which is a complete ordered field distinct from the classical arithmetic, and which is used to reveal certain structural similarities in the classical and exponential calculi. When applied to problems of change and accumulation, classical arithmetic leads to the classical calculus; however, it seems that no one had conceived the idea of using nonclassical arithmetics to construct new systems of calculus. Indeed, before 1967, apparently no one had used a nonclassical arithmetic for any purpose, it having been long believed that there is no distinctive value in the nonclassical arithmetics, since they are all structurally equivalent to classical arithmetic. (Nonclassical arithmetic should be distinguished from the nonstandard arithmetic developed by the logicians.)

Chapter 4 contains graphical interpretations of the exponential calculus; Chapter 5 includes a variety of heuristic principles for selecting appropriate gradients, derivatives, averages, and integrals; Chapter 6 includes a non-Cartesian (analytic) geometry, which is a nonlinear model for plane Euclidean geometry.

Chapter 7 has discussions of exponential vectors, most of which are curvilinear (unlike classical vectors which are all rectilinear), and exponential centroids, which turn up again in Chapter 8 on least-squares methods. The exponential method of least squares provides what is probably the first suitable rationale for the logarithmic-transformation technique that is commonly used for fitting exponential curves to data. Chapter 9 contains, among other things, brief discussions of the percentage derivative and exponential complex-numbers. Bibliographic references and sundry comments have been placed in the NOTES at the ends of the sections. A list of symbols and a detailed index have been provided at the end of the book.

Since this book is intended for a wide audience, I have adopted an expository style that calls for some explanation.

> I have assumed that the reader has a working knowledge of the rudiments of the classical calculus.

> Following the policy of Copernicus, "I shall expound many things differently from my predecessors, although with their aid, for it was they who first opened the road of inquiry into these things."

> I have tried to honor Karl Menger's fine dictum: Where it makes no difference what is said, say nothing. Nevertheless, wherever possible I have set forth the intuitive background to all formal concepts, and I have included a variety of quotations and references, for I sincerely believe that, lacking sufficient historical perspective, few mathematicians would have pursued the analogy that was perceived in July of 1967.

> Holding that mathematics is a reality only in the world of the imagination, I have provided simple but precise definitions, without which nothing is adequately conveyed.

> I have excluded proofs, because what matters here are the results, most of which can be proved in a straightforward way. Nonetheless, the

reader is urged to read critically, for there are no unimportant details in an unfamiliar terrain, where truth and falsity have the same shape.

For their encouragement and help I wish to thank my wife, Jane; my parents; and my friend, teacher, and colleague, Robert Katz, a scientific explorer of extraordinary vision.

Criticisms and suggestions are cordially invited.

Michael Grossman

University of Lowell
Department of Mathematics
Lowell, Massachusetts 01854

December 18, 1979

C O N T E N T S

PRELIMINARIES

In this book the word number means real number. The letter R stands for the set of all numbers, and the symbol R_+ stands for the set of all positive numbers.

If $r < s$, then the interval $[r,s]$ is the set of all numbers x such that $r \leq x \leq s$. (Only such intervals are used here.) The interior of $[r,s]$ consists of all numbers x such that $r < x < s$.

An arithmetic partition of an interval $[r,s]$ is any arithmetic progression whose first term is r and last term is s. An arithmetic partition that has exactly n terms is said to be n-fold.

A point is any ordered pair of numbers, each of which is called a coordinate of the point. A function is a set of points, each distinct two of which have distinct first coordinates.

The arguments of a function are the first coordinates of its points; the domain of a function is the set of all its arguments. A function whose domain is R is said to be on R. A function is also said to be defined at each of its arguments.

The values of a function are the second coordinates of its points; the range of a function is the set of all its values.

If every two distinct points of a function f have distinct second coordinates, then f is one-to-one and its inverse is the one-to-one function consisting of all points (y,x) for which (x,y) is a point of f.

A positive point is any point whose second coordinate is positive. A positive function is any function whose points are all positive.

A discrete function is any function that has only a finite number of arguments.

The function exp is on R and assigns to each number x the number e^x, where e is the base of the natural logarithm function, ln. The function ln is the inverse of exp.

0

CHAPTER 1

The Classical Calculus

1.1 INTRODUCTION

> "The calculus was the first achievement
> of modern mathematics, and it is dif-
> ficult to overestimate its importance.
> I think it defines more unequivocally
> than anything else the inception of
> modern mathematics; and the system of
> mathematical analysis, which is its
> logical development, still constitutes
> the greatest technical advance in
> exact thinking."
>
> John von Neumann[1]

The distinguished mathematician was referring, of
course, to the classical calculus, the system of differential and integral calculus developed by Leibniz, Newton,
and many predecessors, including Eudoxus, Archimedes,
Kepler, Cavalieri, Fermat, Wallis, and Barrow.[2]

In the classical calculus, differences are used to
measure changes (or deviations) in arguments and in values,
and sums are used to accumulate (or combine) arguments and
values. It is fitting, therefore, that Leibniz often referred to the classical calculus as a "calculus of differences and sums."[3] Indeed, Carl B. Boyer pointed out that
"...Leibniz looked upon the operation of finding 'differences' as fundamental in his...calculus."[4]

N O T E S

1. Von Neumann's remark appeared originally in The Works
of the Mind (Chicago: University of Chicago Press, 1947), an
anthology of interesting articles edited by R. B. Heywood.

1

2. Although the term "classical analysis" is often used, the term "classical calculus" has appeared rarely, presumably because there was only one calculus prior to the discovery of the non-Newtonian calculi.

3. In 1696, Leibniz wrote: "I have brought matters so far with my infinitesimal calculus of differences and sums that many problems can now be solved in mathematical physics which one could not even venture to try before." This appeared in Philosophical Papers and Letters of Gottfried Wilhelm Leibniz, Vol. II, ed. and trans. L. E. Loemker (Chicago: University of Chicago Press, 1956), p.768.

4. Carl B. Boyer, The History of the Calculus and Its Conceptual Development (New York: Dover reprint, 1949), p. 206.

1.2 LINEAR FUNCTIONS

> "...the fundamental idea of Calculus,
> namely the 'local' approximation of
> functions by linear functions."
>
> Jean Dieudonné[1]

In the classical calculus, linear functions are the standard to which other functions are compared.

A linear function is any function u on R such that $u(x) = mx + c$, where m and c are constants.[2]

Each linear function u has the following important property:

> For any intervals $[r_1, s_1]$ and $[r_2, s_2]$,
> if $s_1 - r_1 = s_2 - r_2$,
> then $u(s_1) - u(r_1) = u(s_2) - u(r_2)$;
> that is, equal differences in arguments
> yield equal differences in values.[3]

In particular, the number $u(b) - u(a)$ is the same for any numbers a and b such that $b - a = 1$, a fact which suggests the definition in the next section.

N O T E S

1. Jean Dieudonné, Foundations of Modern Analysis (New

York: Academic Press, 1960), p. 141.

2. In Section 1.4 we shall use the fact that there is precisely one linear function containing any two given points with distinct first coordinates.

3. The linear functions are the only continuous functions on R such that equal differences in arguments yield equal differences in values.

1.3 CLASSICAL SLOPE

The classical slope of a linear function u is the number $u(b) - u(a)$, where a and b are any two numbers such that $b - a = 1$.

Though this definition is seldom used, it is well-known, is extremely simple, and is suggestive of other kinds of slopes, one of which is defined in Section 2.3.[1]

Of course the classical slope of the linear function $u(x) = mx + c$ turns out to be m. (The phrase "turns out" will continue to be used to indicate a result that can be proved.)

N O T E

1. Though this definition of classical slope was used, in effect, by Karl Pearson (The Grammar of Science) and undoubtedly by many others, most, if not all, textbook writers use the "rise over run" definition, which, though not incorrect, fails to suggest the extensions that are required for the non-Newtonian calculi. Indeed, one of our early difficulties arose from our failure to use the simpler definition of classical slope.

1.4 THE CLASSICAL GRADIENT

"The importance of the differential calculus arises from the very nature of the subject, which is the systematic consideration of the [gradients] of functions."

Alfred North Whitehead[1]

The differential branch of the classical calculus is rooted in the concept of the average rate of change, which we prefer to call the classical gradient.[2]

The classical gradient of a function f on an interval [r,s] is denoted by $G_r^S f$ and is defined to be the classical slope of the linear function containing the points (r,f(r)) and (s,f(s)).

It turns out that

$$G_r^S f = \frac{f(s) - f(r)}{s - r},$$

which provides a simple tool for calculating $G_r^S f$. The reader is urged, however, to conceive $G_r^S f$ as the classical slope of the linear function containing (r,f(r)) and (s,f(s)), since that conception lends itself to useful generalizations, one of which appears in Section 2.4.

The operator G is

Additive: $G_r^S(f + g) = G_r^S f + G_r^S g,$

Subtractive: $G_r^S(f - g) = G_r^S f - G_r^S g,$

Homogeneous: $G_r^S(c \cdot f) = c \cdot G_r^S f,$ c constant.

Though these three relationships can be expressed in one equation, we prefer to list them separately.

As expected, the classical gradient of a linear function on any interval is equal to its classical slope.

N O T E S

1. A. N. Whitehead, An Introduction to Mathematics (New York: Henry Holt and Company, 1911).

2. The British term "gradient" is better suited for our purpose, since it can be readily modified by appropriate adjectives. Of course, "gradient" is also used in vector analysis, but that subject does not concern us here.

1.5 The Classical Derivative

> "It was Fermat, primarily, who intro-
> duced the modern idea of the tangent
> to a curve at a given point P. In
> essence, he took a second point Q on
> the curve, found the slope of the
> secant line PQ, and from this, by
> permitting Q to tend toward coinci-
> dence with P, he calculated the slope
> of the tangent. This method rightly
> earned for him the title of inventor
> of the differential calculus."
>
> Boyer & Neugebauer[1]

Let f be a function defined at least on an interval containing the number a in its interior. If the following limit[2] exists, we denote it by [Df](a), call it the classical derivative of f at a, and say that f is classically differentiable at a:

$$\lim_{x \to a} \frac{f(x) - f(a)}{x - a}.$$

The classical derivative of f, denoted by Df, is the function that assigns to each number t the number [Df](t), if it exists.[3]

The operator D is additive, subtractive, and homogeneous. (See page 4 for an indication of the meanings of those three terms.)

The classical derivative of a linear function has a constant value equal to its classical slope. Indeed, only linear functions have classical derivatives that are constant on R. In particular, if u is a constant function on R, then Du is everywhere equal to 0.

The familiar concept of tangent line can be defined in a simple way without using limits or derivatives.[4]

It can be proved that [Df](a) exists if and only if f has a tangent at (a,f(a)); and if [Df](a) does exist, it equals the classical slope of that tangent.

Two functions are <u>tangent</u> <u>at</u> <u>a</u> <u>common</u> <u>point</u> if and
only if they have the same tangent there.

<u>N O T E S</u>

1. The remark is from the article "History of Mathematics,"
which appears in some editions of the <u>Encyclopaedia</u> <u>Britan-</u>
<u>nica</u>, e.g., in the 1970 edition.

2. We do not discuss limits and continuity in this book,
since these concepts are identical in the classical and
exponential calculi. However, for most other non-Newtonian
calculi one must introduce appropriately modified concepts
of limit and continuity.

3. Though Laplace rightly remarked that "Leibniz has en-
riched the differential calculus by a very happy notation,"
the Leibniz 'd' notation is not convenient for our purposes.

4. Let f be a function defined at least on an interval
containing the number a in its interior.
The <u>tangent</u> to f at the point (a,f(a)) is the unique
linear function g, if it exists, which possesses the following two properties.
1) The linear function g contains (a,f(a)).
2) For each linear function u containing (a,f(a)) and
distinct from g, there is a positive number p such
that for every number x in [a - p, a + p] but
distinct from a,

$$|g(x) - f(x)| < |u(x) - f(x)|.$$

(Roughly, the tangent is locally closer to f than
any other linear function.)

The preceding definition has the virtue that it can be
generalized in an important way. See <u>Non-Newtonian</u> <u>Calculus</u>,
page 41.

Robert Katz and I formulated this definition of the tan-
gency concept and conjectured its equivalence to the usual
definition, but it was Charles Rockland who kindly furnished
a proof of the equivalence.

1.6 THE ARITHMETIC AVERAGE

"...averaging processes have proved themselves powerful tools in analysis."

Einar Hille

Since values are combined by addition in the classical calculus, it is to be expected that values are averaged by means of the arithmetic average, which is based on addition.

The <u>arithmetic</u> <u>average</u> <u>of</u> <u>n</u> <u>numbers</u> $v_1, \ldots, v_n$ is the number $(v_1 + \cdots + v_n)/n$.

The <u>arithmetic</u> <u>average</u> <u>of</u> <u>a</u> <u>continuous</u> <u>function</u> <u>f</u> <u>on</u> <u>an</u> <u>interval</u> $[r,s]$ is denoted by $M_r^s f$ and is defined to be the limit of the convergent sequence whose nth term is the arithmetic average of $f(a_1), \ldots, f(a_n)$, where $a_1, \ldots, a_n$ is the n-fold arithmetic partition of $[r,s]$.

The operator M is additive, subtractive, and homogeneous, and is characterized by the following three properties. (This use of the term "characterized" indicates that no other operator possesses all three properties.)

For any interval $[r,s]$ and any constant function $h(x) = c$ on $[r,s]$,

$$M_r^s h = c.$$

For any interval $[r,s]$ and any functions f and g that are continuous on $[r,s]$, if $f(x) \leqq g(x)$ for every number x in $[r,s]$, then

$$M_r^s f \leqq M_r^s g.$$

For any numbers r, s, t such that $r < s < t$, and any function f continuous on $[r,t]$,

$$(s - r) \cdot M_r^s f + (t - s) \cdot M_s^t f = (t - r) \cdot M_r^t f.$$

1.7 THE BASIC THEOREM OF CLASSICAL CALCULUS

"The stronghold that certain themes have on the mind of the scientist helps to explain his commitment to some point of view that may run exactly counter to all

accepted doctrine and to the clear
evidence of the senses."

Gerald Holton[1]

Indeed, for many years we have been guided by the
idea that the "kernel" of the classical calculus is neither
the First nor the Second Fundamental Theorem of Classical
Calculus, but rather a well-known result that we call the
Basic Theorem of Classical Calculus. Let us begin with its
discrete analogue, which is a proposition that concerns discrete functions and appropriately conveys the spirit of the
theorem.

The Discrete Analogue of the
Basic Theorem of Classical Calculus

If h is a discrete function whose arguments
$a_1, \ldots, a_n$ are an arithmetic partition of
$[r,s]$, then the arithmetic average of the
classical gradients of h on the intervals
$[a_{i-1}, a_i]$, $i = 2, \ldots, n$, is equal to the
classical gradient of h on $[r,s]$.

The foregoing suggests the following important
theorem.[2]

The Basic Theorem of Classical Calculus

If Dh is continuous on $[r,s]$, then its arithmetic
average on $[r,s]$ equals the classical gradient of
h on $[r,s]$, that is,

$$M_r^s(Dh) = \frac{h(s) - h(r)}{s - r}.$$

In view of this theorem it is appropriate to say that
the arithmetic average fits naturally into the scheme of
classical calculus.

N O T E S

1. Holton's remark appeared in an article for Historical
Studies in the Physical Sciences, 2nd Annual Volume (1970),
edited by Russell McCormmach and published by the University
of Pennsylvania Press.

2. The Basic Theorem of Classical Calculus is a well-known
variant of the Second Fundamental Theorem of Classical
Calculus, which is stated in Section 1.10. Every non-
Newtonian calculus has its own Basic Theorem that is an extension of a simple algebraic identity and has the following
form, in which the underlined words have a special meaning
depending on the nature of the particular calculus:

> If the derivative of a function is continuous
> on an interval, then its average thereon equals
> the gradient of the function thereon.

This is discussed fully in Non-Newtonian Calculus.

One attractive feature of the Basic Theorem, not shared
by the two Fundamental Theorems, is the existence of a
discrete analogue, which can be understood before the introduction of limits, derivatives, and integrals. Accordingly
we consider the Basic Theorem to be the "kernel" of the
classical calculus.

1.8 THE BASIC PROBLEM OF CLASSICAL CALCULUS

Suppose that the value of a function h is known at
an argument r, and suppose that f, the classical
derivative of h, is continuous and known at each
number in [r,s]. Find h(s).

Solution

By the Basic Theorem of Classical Calculus,

$$M_r^s f = M_r^s (Dh) = \frac{h(s) - h(r)}{s - r}.$$

Solving for h(s), we get

$$h(s) = h(r) + (s - r) \cdot M_r^s f. \quad \square$$

The number $(s - r) \cdot M_r^s f$ in the foregoing solution arises with sufficient frequency to warrant a special name, "the classical integral of f on [r,s]," which is introduced in the next section.

Thus, the Basic Theorem of Classical Calculus, which involves the arithmetic average, classical derivative, and classical gradient, provides an immediate solution to the Basic Problem of Classical Calculus, which, in turn, motivates a definition of the classical integral in terms of the arithmetic average.[1]

<u>N O T E</u>

1. Ideally, the introduction of a new concept should be preceded by a problem whose solution suggests the concept. Alas, that cannot always be done in a natural way.

1.9 THE CLASSICAL INTEGRAL

> "The theory of integration is concerned with finding averages of functions."
>
> Edwin Hewitt[1]

The <u>classical</u> <u>integral</u> of a continuous function f on [r,s] is denoted by $\int_r^s f$ and is defined to be the number $(s - r) \cdot M_r^s f$. We set $\int_r^r f = 0$.

It turns out that $\int_r^s f$ equals the limit of the convergent sequence whose nth term is the sum

$$k_n \cdot f(a_1) + \cdots + k_n \cdot f(a_{n-1}),$$

where $a_1, \ldots, a_n$ is the n-fold arithmetic partition of [r,s], and k_n is the common value of

$$a_2 - a_1, \; a_3 - a_2, \; \ldots, \; a_n - a_{n-1}.$$

The operator $\int$ is additive, subtractive, and homogeneous, and is characterized by the following three properties.

For any interval $[r,s]$ and any constant function $h(x) = c$ on $[r,s]$,

$$\int_r^s h = (s - r) \cdot c.$$

For any interval $[r,s]$ and any functions f and g that are continuous on $[r,s]$, if $f(x) \leqq g(x)$ for every number x in $[r,s]$, then

$$\int_r^s f \leqq \int_r^s g.$$

For any numbers r, s, t such that $r < s < t$, and any function f continuous on $[r,t]$,

$$\int_r^s f + \int_s^t f = \int_r^t f.$$

In Section 5.4 there is an important application of the foregoing characterization.

The classical integral can be characterized in other interesting and useful ways.[2]

N O T E S

1. The remark occurs on page 50 of Hewitt's preliminary edition of <u>Theory of Functions of a Real Variable</u> (New York: Holt, Rinehart, and Winston, 1960). In their <u>Real and Abstract Analysis</u> (New York: Springer-Verlag Inc., 1965), Hewitt and Stromberg asserted that "Integration from one point of view is an averaging process for functions...."
    It is surprising, nevertheless, that averages do not seem to play a central role in modern analysis. For example, only rarely does one encounter the geometric, quadratic, and harmonic averages in integration theory.
    In a special course given at Harvard University in 1958,

Robert Katz defined the classical integral in terms of the arithmetic average, believing that such a procedure is intuitively more satisfying. Subsequently, in his courses at Tufts University, he always used that technique, which found considerable favor with the students and the engineering faculty, especially since the use of the arithmetic average simplifies the presentation of many, but not all, scientific concepts. (For example, see Section 5.5 herein.)

2. Excellent characterizations of the classical integral may be found in Chapters 5 and 6 of Calculus (New York: W. W. Norton, 1973) by Leonard Gillman and Robert H. McDowell.

1.10 THE FUNDAMENTAL THEOREMS OF CLASSICAL CALCULUS

> "Areas by integration had been found, through summations, by earlier mathematicians from Archimedes to Wallis; and differentiations had been carried out by Fermat. It remained for Newton and Gottfried Leibniz to discover the fundamental principle of the calculus — that integrations can be performed far more easily by inverting the process of differentiation."
>
> Boyer & Neugebauer[1]

The classical derivative and integral are 'inversely' related in the sense indicated by the following two theorems.

First Fundamental Theorem of Classical Calculus

If f is continuous on [r,s], and

$$g(x) = \int_r^x f , \quad \text{for every number } x \text{ in } [r,s],$$

then

$$Dg = f, \quad \text{on } [r,s].[2]$$

Second Fundamental Theorem of Classical Calculus

If Dh is continuous on [r,s], then

$$\int_r^s (Dh) = h(s) - h(r).$$

The latter theorem is a simple consequence of the Basic Theorem discussed in Section 1.7.

This concludes our brief discussion of the classical calculus.

N O T E S

1. For a citation see Note 1 to Section 1.5.

2. The following theorem reveals another interesting property of the function $\int_r^x f$.

Let $r < a < s$ and assume f is continuous on $[r,s]$. Then

$$G_a^s \left[\int_r^x f \right] = M_a^s f.$$

CHAPTER 2

The Exponential Calculus

2.1 INTRODUCTION

> "All perception of truth is the
> detection of an analogy."
>
> Thoreau

The initial stimulus for the creation of the exponential calculus was a recognition of the following analogy.

Let h be a discrete function whose arguments a_1, ..., a_n are in arithmetic progression. According to the Discrete Analogue of the Basic Theorem of Classical Calculus (Section 1.7), the arithmetic average of the classical gradients

$$\frac{h(a_i) - h(a_{i-1})}{a_i - a_{i-1}}, \quad i = 2, \ldots, n$$

is equal to the classical gradient

$$\frac{h(a_n) - h(a_1)}{a_n - a_1}.$$

If one assumes, furthermore, that the discrete function h is positive, then it readily follows that the geometric average of the numbers

$$\left[\frac{h(a_i)}{h(a_{i-1})} \right]^{1/(a_i - a_{i-1})}, \quad i = 2, \ldots, n$$

is equal to

$$\left[\frac{h(a_n)}{h(a_1)}\right]^{1/(a_n - a_1)} .$$

We imagined, therefore, that those numbers are gradients of a new kind (exponential gradients), and we conjectured that the foregoing relationship between exponential gradients is the discrete analogue of the basic theorem of a new calculus, the exponential calculus.

The methods of determining changes and accumulations of arguments and values are compared in the following chart.

	classical calculus	exponential calculus
changes of arguments:	differences	differences
changes of values:	differences	ratios
accumulations of arguments:	sums	sums
accumulations of values:	sums	products

Moreover, the operators of the exponential calculus are applied only to positive functions; hence every function in this chapter is positive, except for the natural logarithm function, ln, which is used for special purposes. Lest the reader be disappointed by such a restriction, we hasten to say that an exponential-type calculus for negative-valued functions is discussed in Section 6.10 of Non-Newtonian Calculus.

It is certainly not unusual to measure deviations by ratios rather than by differences. For instance, during the Renaissance many scholars, including Galileo, discussed the following problem:

Two estimates, 10 and 1000, are proposed as the value of a horse. Which estimate, if any, deviates more from the true value of 100?

The scholars who maintained that the deviations should be measured by differences concluded that the estimate of 10 was closer to the true value. However, Galileo eventually maintained that the deviations should be measured by ratios and that accordingly the two estimates deviated equally from the true value.

Four hundred years later the controversy about ratios and differences was still alive, as indicated by the following remark by William H. Kruskal.[1]

> "In recent discussions of possible relationships between cigarette smoking and lung cancer, controversy arose over whether ratios or differences of mortality rates were of central importance. The choice may lead to quite different conclusions."

It is also worth noting that the ear's subjective comparison of loudness is apparently achieved by sensing ratios, not differences, of sound levels.[2]

Now let us consider the following problem. At time r, a man invests f(r) dollars with a promoter who guarantees that at a certain subsequent time s, the value of the investment will be f(s) dollars. In event that the investor should desire to withdraw at any other time t, it was agreed that the value of the investment increases continuously and 'uniformly.' The problem is this: How much, f(t), would the investor be entitled to at time t? We shall give two reasonable solutions; there is no unique solution to this problem.

Solution 1. Since it was agreed that the value of the investment increases 'uniformly,' we may reasonably assume that it increases by equal amounts in equal times. Since, furthermore, the value of the investment increases continuously, it can be proved that the growth must be linear and, in fact, that

$$f(t) = f(r) + \left[\frac{f(s) - f(r)}{s - r} \right] (t - r).$$

Notice that the expression within the brackets represents
the classical gradient of f on the time-interval [r,s].
Solution 2. Since the value of the investment increases
'uniformly,' we may reasonably assume that it increases by
equal ratios in equal times. Since, furthermore, the value
of the investment increases continuously, it can be proved
that the growth must be exponential and, in fact, that

$$
f(t) = f(r) \left\{ \left[\frac{f(s)}{f(r)} \right]^{\frac{1}{s-r}} \right\}^{t-r} .
$$

We shall see in Section 2.4 that the expression within the
braces is of fundamental importance in the exponential
calculus.

The operators of the exponential calculus will be interpreted graphically in Chapter 4, and heuristic principles
for their application will be given in Chapter 5.

We find it convenient to use the prefix "*-" to stand
for "exponential" or "exponentially," whichever is grammati-
cally appropriate.

N O T E S

1. The remark was made in Kruskal's article "Statistics:
The Field," which appears in Volume 15 of the International
Encyclopedia of the Social Sciences, edited by David L.
Sills and published by The Macmillan Company and The Free
Press in 1968.

2. See, e.g., William C. Vergara, Science, the Never-Ending
Quest (Harper & Row, 1965), p. 132.

2.2 EXPONENTIAL FUNCTIONS

In the *-calculus, exponential functions are the
standard to which other functions are compared.

By an _exponential_ _function_ we mean any function u on
R such that u(x) = exp(mx + c), where m and c are constants.
According to that definition, every exponential function is
positive and every positive constant function on R is
exponential.[1]

If p is a positive constant, then the functions
v(x) = p^{mx+c} are also exponential, since they are expressible
in the required form.

Each exponential function u has the following important property:

> For any intervals $[r_1,s_1]$ and $[r_2,s_2]$,
> if $s_1 - r_1 = s_2 - r_2$,
> then $u(s_1)/u(r_1) = u(s_2)/u(r_2)$;
> that is, equal differences in arguments
> yield equal ratios of values.[2]

In particular, the number u(b)/u(a) is the same for any numbers a and b such that b - a = 1, a fact which suggests the
definition in the next section.

N O T E S

1. The following fact will be required in Section 2.4:
there is precisely one exponential function containing any
two given positive points with different first coordinates.

2. The exponential functions are the only positive functions that are continuous on R and have the property that
equal differences in arguments yield equal ratios of values.

2.3 EXPONENTIAL SLOPE

> "The language of analysis...and its
> notations...are so many germs of
> new calculi."
>
>   Laplace

The *-slope of an exponential function u is the positive number u(b)/u(a), where a and b are any two numbers
such that b - a = 1. (The reader may wish to compare that

with the definition of classical slope in Section 1.3.)

The *-slopes of the exponential functions u(x) = exp(mx + c) and v(x) = p^{mx+c} (p a positive constant) turn out to be e^m and p^m, respectively.

2.4 THE EXPONENTIAL GRADIENT

"To new concepts correspond,
necessarily, new symbols."

Gauss

The differential branch of the *-calculus is rooted in the concept of the *-gradient.

The *-gradient of a positive function f on an interval [r,s] is denoted by $\overset{*}{G}{}^{s}_{r}f$ and is defined to be the *-slope of the exponential function containing (r,f(r)) and (s,f(s)). It turns out that

$$\overset{*}{G}{}^{s}_{r}f = \left[\frac{f(s)}{f(r)} \right]^{\frac{1}{s-r}},$$

thus furnishing a way of calculating $\overset{*}{G}{}^{s}_{r}f$. Nevertheless, the reader is urged to conceive $\overset{*}{G}{}^{s}_{r}f$ in the manner defined above.[1]

The operator $\overset{*}{G}$ is

Multiplicative: $\overset{*}{G}{}^{s}_{r}(f \cdot g) = \overset{*}{G}{}^{s}_{r}f \cdot \overset{*}{G}{}^{s}_{r}g$,

Divisional: $\overset{*}{G}{}^{s}_{r}(f/g) = \overset{*}{G}{}^{s}_{r}f / \overset{*}{G}{}^{s}_{r}g$,

Involutional: $\overset{*}{G}{}^{s}_{r}(f^{c}) = (\overset{*}{G}{}^{s}_{r}f)^{c}$, c constant.

Of course the *-gradient of an exponential function on any interval is equal to its *-slope.

The *-gradient is related to the classical gradient as follows:

$$\overset{*}{G}{}^{s}_{r}f = \exp \left[G^{s}_{r}(\ln f) \right].$$

In Section 9.2 we shall discuss a connection between the *-gradient and the so-called compound growth rate.

When r = s, the expression for the *-gradient yields the indeterminate form 1^∞, in contrast to the indeterminate form 0/0 yielded by the expression for the classical gradient.

N O T E

1. Some reviewers of <u>Non-Newtonian</u> Calculus have unwittingly vitiated our indications that we do NOT define gradients by formulas. Nothing would be more unenlightening than a discussion that began thus:

"The *-gradient of a positive function f on [r,s] is defined to be

$$\left[\frac{f(s)}{f(r)}\right]^{1/(s-r)} \quad ."$$

2.5 THE EXPONENTIAL DERIVATIVE

Let f be a positive function defined at least on an interval containing the number a in its interior. If the following limit exists and is positive,[1] we denote it by $[\overset{*}{D}f](a)$, call it the *-derivative of f at a, and say that f is *-differentiable at a:

$$\lim_{x \to a}\left[\frac{f(x)}{f(a)}\right]^{\frac{1}{x-a}} .$$

It can be proved that [Df](a) and $[\overset{*}{D}f](a)$ coexist; that is, if either exists then so does the other. Moreover, if they do exist, then $[\overset{*}{D}f](a)$ equals the *-slope of the unique exponential function that is tangent to f at (a, f(a)).

The following relationship is clearly similar to the relationship between gradients stated in the preceding section:

$$[\overset{*}{D}f](a) = \exp\left\{[D(\ln f)](a)\right\}. \quad \text{(See Note 2.)}$$

The *-derivative of f, denoted by $\overset{*}{D}f$, is the function that assigns to each number t the number $[\overset{*}{D}f](t)$, if it exists.

The *-derivative of an exponential function has a constant value equal to its *-slope. Indeed, only exponential functions have *-derivatives that are constant on R. In particular, if u is a positive constant function on R, then $\overset{*}{D}u$ is everywhere equal to 1.[3]

The operator $\overset{*}{D}$ is multiplicative, divisional, and involutional. (See page 19 for an indication of the meanings of those three terms.)

It is worth noting that the function w(x) = exp[exp(x)] equals its *-derivative, that is, $\overset{*}{D}w = w$.

The relationship between the nth *-derivative and the nth classical derivative follows the familiar pattern:

$$[\overset{*}{D}^n f](a) = \exp\left\{[D^n(\ln f)](a)\right\}.$$

The second *-derivative of the function $\exp(-x^2)$ turns out to be the constant $1/e^2$, a fact that may shed some additional insight into that function, which plays an important role in probability and statistics.

N O T E S

1. It is possible for the limit to be 0; that happens, for example, if $f(x) = \exp(-x^{1/3})$ and a = 0. The reason for requiring the limit to be positive is best explained in the context of the general theory of the non-Newtonian calculi and is given in Section 6.10 of Non-Newtonian Calculus.

2. Hence $[\overset{*}{D}f](a) = \exp\left\{[Df](a)/f(a)\right\}$. The expression within the braces is the well-known logarithmic derivative.

3. Therefore $\overset{*}{D}(1 + 1) \neq \overset{*}{D}(1) + \overset{*}{D}(1)$, and so the operator $\overset{*}{D}$ is not additive. The reader may enjoy finding a formula for $\overset{*}{D}(f + g)$.

2.6 The Geometric Average

Since values are combined by multiplication in the exponential calculus, it is to be expected that values would be averaged by means of the geometric average, which is based on multiplication.

The geometric average of n positive numbers w_1, ..., w_n is the positive number $(w_1 w_2 \cdots w_n)^{1/n}$. [1]

The geometric average of a continuous positive function f on an interval [r,s] is denoted by $\overset{*}{M}{}_r^s f$ and is defined to be the positive limit of the convergent sequence whose nth term is the geometric average of $f(a_1)$, ..., $f(a_n)$, where a_1, ..., a_n is the n-fold arithmetic partition of [r,s].

The operator $\overset{*}{M}$ is multiplicative, divisional, and involutional, and is characterized by the following three properties.

For any interval [r,s] and any positive constant function $h(x) = p$ on [r,s],

$$\overset{*}{M}{}_r^s h = p.$$

For any interval [r,s] and any positive functions f and g that are continuous on [r,s], if $f(x) \leqq g(x)$ for every number x in [r,s], then

$$\overset{*}{M}{}_r^s f \leqq \overset{*}{M}{}_r^s g.$$

For any numbers r, s, t such that $r < s < t$, and any positive function f continuous on [r,t],

$$[\overset{*}{M}{}_r^s f]^{s-r} \cdot [\overset{*}{M}{}_s^t f]^{t-s} = [\overset{*}{M}{}_r^t f]^{t-r}.$$

The geometric and arithmetic averages are related as follows:

$$\overset{*}{M}{}_r^s f = \exp\left\{ M_r^s(\ln f) \right\}.$$

N O T E

1. For our purposes the expression "exponential average" is more suitable than "geometric average." However, we yield to customary usage.

2.7 THE BASIC THEOREM OF EXPONENTIAL CALCULUS

As in the discussion of the Basic Theorem of Classical Calculus (Section 1.7), we begin with a discrete analogue.

The Discrete Analogue of the
Basic Theorem of *-Calculus

If h is a discrete positive function whose arguments a_1, ..., a_n are an arithmetic partition of [r,s], then the geometric average of the *-gradients of h on the intervals $[a_{i-1}, a_i]$, i = 2, ..., n, is equal to the *-gradient of h on [r,s].

The foregoing result suggests the following important theorem.

The Basic Theorem of *-Calculus

If $\overset{*}{D}h$ is continuous on [r,s], then its geometric average on [r,s] equals the *-gradient of h on [r,s], that is,

$$\overset{*}{M}{}_{r}^{s}(\overset{*}{D}h) = \left[\frac{h(s)}{h(r)} \right]^{\frac{1}{s-r}} .$$

In view of this theorem it is appropriate to say that the geometric average fits naturally into the scheme of *-calculus.

2.8 THE BASIC PROBLEM OF EXPONENTIAL CALCULUS

Suppose that the value of a positive function h is known at an argument r, and suppose that f, the *-derivative of h, is continuous and known at each number in [r,s]. Find h(s).

Solution

By the Basic Theorem of *-Calculus,

$$\overset{*s}{M_r}f = \overset{*s}{M_r}(\overset{*}{D}h) = \left[\frac{h(s)}{h(r)} \right]^{\frac{1}{s-r}}.$$

Solving for h(s), we get

$$h(s) = h(r) \cdot [\overset{*s}{M_r}f]^{s-r} . \quad \Box$$

The number $[\overset{*s}{M_r}f]^{s-r}$ that appears in the foregoing solution will arise with sufficient frequency to warrant a special name, "the *-integral of f on [r,s]," which is introduced in the next section.

Thus, the Basic Theorem of *-Calculus, which involves the geometric average, *-derivative, and *-gradient, provides an immediate solution to the Basic Problem of *-Calculus, which, in turn, motivates our definition of the *-integral in terms of the geometric average. (Every non-Newtonian calculus has a Basic Problem, whose solution via its Basic Theorem motivates a definition of the integral in that calculus.)

2.9 THE EXPONENTIAL INTEGRAL

The *-integral of a continuous positive function f on [r,s] is denoted by $\displaystyle\int_r^{*s} f$ and is defined to be the positive number $[M_r^s f]^{s-r}$. We set $\int_r^{*r} f = 1$.

It turns out that $\int_r^{*s} f$ equals the positive limit of the convergent sequence whose nth term is the product

$$[f(a_1)]^{k_n} \cdot [f(a_2)]^{k_n} \cdots [f(a_{n-1})]^{k_n} ,$$

where $a_1, \ldots, a_n$ is the n-fold arithmetic partition of $[r,s]$, and k_n is the common value of

$$a_2 - a_1, a_3 - a_2, \ldots, a_n - a_{n-1}.$$

The operator $\int^*$ is multiplicative, divisional, and involutional, and is characterized by the following three properties.

For any interval $[r,s]$ and any positive constant function $h(x) = p$ on $[r,s]$,

$$\int_r^{*s} h = p^{s-r} .$$

For any interval $[r,s]$ and any positive functions f and g that are continuous on $[r,s]$, if $f(x) \leq g(x)$ for every number x in $[r,s]$, then

$$\int_r^{*s} f \leq \int_r^{*s} g .$$

For any numbers r, s, t such that $r < s < t$, and any positive function f continuous on $[r,t]$,

$$\int_r^{*s} f \cdot \int_s^{*t} f = \int_r^{*t} f .$$

In Section 5.4, there is an application of the foregoing characterization.

The *-integral and classical integral are related
thus:

$$\int_{r}^{*s} f = \exp \left\{ \int_{r}^{s} (\ln f) \right\} .$$

2.10 The Fundamental Theorems Of Exponential Calculus

The *-derivative and integral are 'inversely' related in the sense indicated by the following two theorems, the
second of which is a simple consequence of the Basic Theorem
discussed in Section 2.7.

First Fundamental Theorem of *-Calculus

If f is positive and continuous on [r,s], and

$$g(x) = \int_{r}^{*x} f , \text{ for every number x in } [r,s],$$

then

$$\overset{*}{D}g = f, \text{ on } [r,s].[1]$$

Second Fundamental Theorem of *-Calculus

If $\overset{*}{D}h$ is continuous on [r,s], then

$$\int_{r}^{*s} (\overset{*}{D}h) = h(s)/h(r).$$

Just as the Second Fundamental Theorem of Classical
Calculus is useful for evaluating classical integrals, the
Second Fundamental Theorem of *-Calculus is useful for evaluating *-integrals. For example, let $f(x) = \exp(1/x)$ and
$h(x) = x$, for $x > 0$. Then $f = \overset{*}{D}h$, and so

$$\int_{3}^{*5} f = \int_{3}^{*5} \overset{*}{(Dh)} = h(5)/h(3) = 5/3.$$

N O T E

1. The following theorem reveals another interesting property of the function $\int_{r}^{*x} f$.

Let $r < a < s$ and assume f is positive and continuous on $[r,s]$. Then

$$\overset{*s}{G_a} \left[\int_{r}^{*x} f \right] = \overset{*s}{M_a} f.$$

2.11 SUMMARY OF RELATIONSHIPS TO THE CLASSICAL CALCULUS

The operators of the *-calculus are uniformly related to the corresponding operators of the classical calculus:

(1) $\quad \overset{*s}{G_r} f \quad = \quad \exp\left\{ G_r^s(\ln f) \right\}$

(2) $\quad \overset{*}{D} f \quad = \quad \exp\left\{ D(\ln f) \right\}$

(3) $\quad \overset{*s}{M_r} f \quad = \quad \exp\left\{ M_r^s(\ln f) \right\}$

(4) $\quad \int_{r}^{*s} f = \exp\left\{ \int_{r}^{s}(\ln f) \right\}$

Relationship (3) is, of course, well-known. An expression of the form $\exp\left\{ \int_{r}^{s}(\ln f) \right\}$ occurs in Inequalities by Hardy, Littlewood, and Pólya (Cambridge University Press, 1952), but the authors do not identify it as an integral; however, as far as we can determine, that book contains no expressions of the form $\exp\left\{ D(\ln f) \right\}$.

Indeed, we have never seen the expressions on the right sides of (1) and (2) in the literature. And we have never encountered a suggestion that there might be a calculus distinct from the classical. (See, however, Section 9.4.)

Of course, it is also possible to express each classical operator in terms of the corresponding *-operator. For example,

$$G_r^s f = \ln \left\{ \overset{*}{G}_r^s [\exp(f)] \right\}.$$

These relationships suggest that for each theorem in classical calculus there is a corresponding theorem in *-calculus, and conversely. For instance, here is a Mean Value Theorem of *-Calculus:

> If a positive function f is continuous on [r,s] and *-differentiable everywhere between r and s, then between r and s there is a number at which the *-derivative of f equals the *-gradient of f on [r,s].

If the reader suspects that operators of other non-Newtonian calculi can be obtained by replacing exp and ln in (1) to (4) by other suitable pairs of inversely-related functions, he or she is correct. But apt formulations and interpretations of such calculi are not simple matters. Furthermore, there are infinitely-many non-Newtonian calculi that cannot be so obtained. (Such issues are discussed in detail in Non-Newtonian Calculus.) Nevertheless the following remark attributed to Hilbert is most appropriate here:

> "The art of doing mathematics consists
> in finding that special case which
> contains all the germs of generality."

For us the special case was the *-calculus.

CHAPTER 3

Exponential Arithmetic

3.1 Introduction

> "The laws of number...are not the
> laws of nature..., they are laws
> of the laws of nature."
>
> Frege

Classical calculus and Cartesian analytic geometry
are based on classical arithmetic, which is usually called
the real number system. But it was the use of nonclassical
arithmetics that led to a general theory of the non-Newtonian
calculi, to the development of non-Cartesian analytic geometries, and to the conception of new kinds of vectors,
centroids, least-squares methods, and complex numbers.
Furthermore, nonclassical arithmetics may also be useful in
devising new systems of measurement that will yield simpler
physical laws. This was clearly recognized by Norman Robert
Campbell, a pioneer in the theory of measurement:

> "...we must recognize the possibility
> that a system of measurement may be
> arbitrary otherwise than in the choice
> of unit; there may be arbitrariness in
> the choice of the process of addition."[1]

In this chapter we shall describe one nonclassical
arithmetic, exponential arithmetic, which will be used
throughout the remainder of the book.

N O T E

1. The quotation is from Campbell's Foundations of Science
(Dover reprint, 1957), p. 292. A very remarkable book.

3.2 CLASSICAL ARITHMETIC

Classical arithmetic has been used for centuries but
was not established on a sound axiomatic basis until the
latter part of the nineteenth century. However, the details
of such a treatment are not essential here.[1]

Informally, <u>classical</u> <u>arithmetic</u> is a system consisting of a set R, for which there are four operations +, -, ×,
/ and an ordering relation <, all subject to certain axioms
commonly referred to as the complete-ordered-field axioms.
The members of R are called numbers, and the set R is called
the <u>realm</u> of classical arithmetic.[2]

N O T E S

1. A new axiomatic treatment of classical arithmetic and a
novel approach to basic logic and set theory are presented
in detail in Robert Katz's <u>Axiomatic</u> <u>Analysis</u>, which was
prepared under the general editorship of David V. Widder and
may be obtained from the publisher of the present book. The
first full axiomatization of classical arithmetic was probably given by Hilbert in his article "Uber den Zahlbegriff,"
1900. The term "classical arithmetic" was used by William
and Martha Kneale in <u>The</u> <u>Development</u> <u>of</u> <u>Logic</u> (Oxford University Press, 1962).

2. The reader is reminded that in this book the word "number" means real number. The term "realm" was introduced in
<u>Non-Newtonian</u> <u>Calculus</u>.

3.3 EXPONENTIAL ARITHMETIC

By ⋆- <u>arithmetic</u> we mean the system consisting of
the set R_+ of all positive numbers, the usual ordering rela-
tion <, and four operations $\overset{*}{+}$, $\overset{*}{-}$, $\overset{*}{\times}$, $\overset{*}{/}$ defined for R_+ as
follows:

<u>⋆-addition</u> $a \overset{*}{+} b = \exp(\ln a + \ln b)$

<u>⋆-subtraction</u> $a \overset{*}{-} b = \exp(\ln a - \ln b)$

*-multiplication $\quad a \overset{*}{\times} b = \exp(\ln a \times \ln b)$

*-division $\qquad\;\; a \overset{*}{/} b = \exp(\ln a / \ln b), \; b \neq 1.$

The reader should note that $a \overset{*}{+} b = ab$ and $a \overset{*}{-} b = a/b$. The realm of *-arithmetic is R_+.

The fact that *-arithmetic satisfies all the complete-ordered-field axioms has an important practical implication: the rules for handling *-arithmetic are exactly the same as the rules for handling classical arithmetic. For example:

$$a \overset{*}{+} b = b \overset{*}{+} a$$

$$a \overset{*}{\times} b = b \overset{*}{\times} a$$

$$a \overset{*}{\times} (b \overset{*}{+} c) = (a \overset{*}{\times} b) \overset{*}{+} (a \overset{*}{\times} c)$$

For each number r, we set $\overset{*}{r} = e^r$. For example, $\overset{*}{0} = e^0 = 1$ and $\overset{*}{1} = e^1 = e$.

Since

$$y \overset{*}{+} \overset{*}{0} = y \text{ and } y \overset{*}{\times} \overset{*}{1} = y$$

for each number y in R_+, we see that $\overset{*}{0}$ and $\overset{*}{1}$ are the "zero" and "one" in *-arithmetic.[1]

The *-integers are the numbers $\overset{*}{n}$, where n is an arbitrary integer; if $\overset{*}{0} < \overset{*}{n}$, then

$$\overset{*}{n} = \underbrace{\overset{*}{1} \overset{*}{+} \cdots \overset{*}{+} \overset{*}{1}}_{n \text{ terms}}.$$

The *-positive numbers are the numbers in R_+ greater than $\overset{*}{0}$, and the *-negative numbers are the numbers in R_+ less than $\overset{*}{0}$.

For each number p in R_+, we make the following definitions:

$$\overset{*}{-}p = \overset{*}{0} \overset{*}{-} p;$$

$$p^{\overset{*}{2}} = p \overset{*}{\times} p; \quad \text{(See Note 2.)}$$

$$\overset{\star}{|}p\overset{\star}{|} = \begin{cases} p & \text{if } p \geq \overset{\star}{0} \\ \overset{\star}{-}p & \text{if } p < \overset{\star}{0}; \end{cases}$$

if $p \geq \overset{\star}{0}$, then $\overset{\star}{\sqrt{}}p$ is the unique number s in R_+

such that $s \geq \overset{\star}{0}$ and $s^{\overset{\star}{2}} = p$.

It turns out that

$$\overset{\star}{-}(\overset{\star}{-}p) = p;$$

$$(\overset{\star}{\sqrt{}}p)^{\overset{\star}{2}} = p, \quad \text{if } p \geq \overset{\star}{0};$$

$$\overset{\star}{\sqrt{}}p^{\overset{\star}{2}} = \overset{\star}{|}p\overset{\star}{|}.$$

Also,

$$\overset{\star}{-}p = \exp\{-\ln p\} = 1/p;$$

$$p^{\overset{\star}{2}} = \exp\{(\ln p)^2\};$$

$$\overset{\star}{|}p\overset{\star}{|} = \exp\{|\ln p|\};$$

$$\overset{\star}{\sqrt{}}p = \exp\{\sqrt{\ln p}\}, \quad \text{if } p \geq \overset{\star}{0}.$$

The following comparisons show that the role of the geometric average in ∗-arithmetic is similar to the role of the arithmetic average in classical arithmetic.

Let v be the arithmetic average of n numbers v_1, ..., v_n, and let w be the geometric average of n positive numbers w_1, ..., w_n. Then v, which equals

$$(v_1 + \cdots + v_n)/n,$$

is the unique number such that

$$\underbrace{v + \cdots + v}_{n \text{ terms}} = v_1 + \cdots + v_n;$$

and w, which equals

$$(w_1 \overset{*}{+} \cdots \overset{*}{+} w_n) \overset{**}{/} n,$$

is the unique positive number such that

$$\underbrace{w \overset{*}{+} \cdots \overset{*}{+} w}_{n \text{ terms}} = w_1 \overset{*}{+} \cdots \overset{*}{+} w_n.$$

(Thus, it is appropriate to say that the arithmetic and geometric averages are the 'natural' averages of classical and *-arithmetics, respectively.)

Furthermore,

(1) $(v - v_1) + \cdots + (v - v_n) = 0;$

(2) the expression

$$\sqrt{(x - v_1)^2 + \cdots + (x - v_n)^2},$$

where x is unrestricted in R,
is a minimum when and only when x = v;

(3) $(w \overset{*}{-} w_1) \overset{*}{+} \cdots \overset{*}{+} (w \overset{*}{-} w_n) = \overset{*}{0};$

(4) the expression

$$\overset{*}{\sqrt{(x \overset{*}{-} w_1)^{\overset{*}{2}} \overset{*}{+} \cdots \overset{*}{+} (x \overset{*}{-} w_n)^{\overset{*}{2}}}},$$

where x is unrestricted in R_+,
is a minimum when and only when x = w.

It is convenient to conceive the radical expression in (2) above as representing the 'classical distance' from x to $v_1, \ldots, v_n$. (For n = 1, the expression reduces to $|x - v_1|$, which is the usual distance from x to v_1.)

34

Accordingly, one may say that the arithmetic average of
v_1, ..., v_n is the number that is 'classically closest' to
v_1, ..., v_n.

Similarly, it is convenient to conceive the radical
expression in (4) above as representing the '$*$-distance'
from x to w_1, ..., w_n. (For n = 1, the '$*$-distance' equals
$\overset{*}{|}x \overset{*}{-} w_1\overset{*}{|}$.) Thus one may say that the geometric average of
w_1, ..., w_n is the positive number that is '$*$-closest' to
w_1, ..., w_n.

The following two facts show that the geometric progressions play the same role in $*$-arithmetic as the arithmetic progressions do in classical arithmetic.

An arithmetic progression is a finite sequence
of numbers v_1, ..., v_n such that $v_i - v_{i-1}$ is
the same for every integer i from 2 to n.

A geometric progression is a finite sequence of
positive numbers w_1, ..., w_n such that
$w_i \overset{*}{-} w_{i-1}$ is the same for every integer i from
2 to n.

It should be quite clear that every concept in classical arithmetic has a counterpart in $*$-arithmetic.

Obviously $*$-arithmetic is especially useful in situations where products and ratios provide the natural methods
of combining and comparing magnitudes. Although $*$-arithmetic
applies only to positive numbers, it is easy to construct a
$*$-type arithmetic that applies to negative numbers.[3]

N O T E S

1. The stipulation in Section 2.9 that $\int_r^{\overset{*}{}r} f = \overset{*}{0}$ clearly

parallels the stipulation in Section 1.9 that $\int_r^r f = 0$.

2. Since $\overset{*}{2} = e^2$, there is a slight risk that the reader will take $p^{\overset{*}{2}}$ to be $p^{(e^2)}$. We wish to stress that $p^{\overset{*}{2}}$ is defined to be $p \overset{*}{\times} p$, which equals $\exp\{(\ln p)^2\}$.

3. See <u>Non-Newtonian Calculus</u>, Section 5.4.

3.4 COMPARISON OF THE CLASSICAL AND EXPONENTIAL CALCULI

First let it be noted that the exponential functions $u(x) = \exp(mx + c)$, which are the standard of comparison in the *-calculus, may be expressed as $u(x) = (\overset{*}{m} \overset{*}{\times} \overset{*}{x}) \overset{*}{+} \overset{*}{c}$, thus exhibiting their similarity to the linear functions, which are the standard of comparison in the classical calculus.

By using *-arithmetic to express the operators of the *-calculus, we shall reveal some similarities between the classical and *-operators.

Let $t = s - r$, where $r < s$.

<u>Gradients</u>

$$G_r^s f = [f(s) - f(r)]/t$$

$$\overset{*}{G}_r^s f = [f(s) \overset{*}{-} f(r)]\overset{**}{/}t$$

<u>Integrals</u>

$$\int_r^s f = t \times M_r^s f$$

$$\overset{*}{\int_r^s} f = \overset{*}{t} \overset{*}{\times} \overset{*}{M}_r^s f$$

<u>Averages</u>

If $a_1, \ldots, a_n$ is the n-fold arithmetic partition of $[r,s]$, then

$$M_r^s f = \lim_{n \to \infty} \left\{ [f(a_1) + \cdots + f(a_n)] / n \right\}$$

$$\overset{*}{M}_r^s f = \lim_{n \to \infty} \left\{ [f(a_1) \overset{*}{+} \cdots \overset{*}{+} f(a_n)] \overset{*}{/} \overset{*}{n} \right\}$$

It was noted in Section 2.4 that the *-gradient is multiplicative, divisional, and involutional:

$$\overset{*s}{G_r}(f \cdot g) = \overset{*s}{G_r}f \cdot \overset{*s}{G_r}g,$$

$$\overset{*s}{G_r}(f/g) = \overset{*s}{G_r}f / \overset{*s}{G_r}g,$$

$$\overset{*s}{G_r}(f^c) = (\overset{*s}{G_r}f)^c, \quad c \text{ constant.}$$

By using *-arithmetic to re-express those three properties, we find that they are actually conditions of additivity, subtractivity, and homogeneity within *-arithmetic:

$$\overset{*s}{G_r}(f \overset{*}{+} g) = \overset{*s}{G_r}f \overset{*}{+} \overset{*s}{G_r}g,$$

$$\overset{*s}{G_r}(f \overset{*}{-} g) = \overset{*s}{G_r}f \overset{*}{-} \overset{*s}{G_r}g,$$

$$\overset{*s}{G_r}(c \overset{*}{\times} f) = c \overset{*}{\times} \overset{*s}{G_r}f, \quad c \text{ constant.}$$

The corresponding properties of the operators $\overset{*}{D}$, $\overset{*}{M}$, and $\overset{*}{\int}$ may be re-expressed similarly.

3.5 Arithmetics And Calculi

An <u>arithmetic</u> is any system that satisfies the complete-ordered-field axioms and has a realm that is a subset of R. There are infinitely-many arithmetics, all of which are isomorphic, i.e., structurally equivalent. Nevertheless the fact that two systems are isomorphic does NOT preclude their separate uses.[1]

Each ordered pair of arithmetics gives rise to a calculus by a judicious use of the first arithmetic for function arguments and the second arithmetic for function values. The following chart indicates the four calculi obtainable by using the classical and exponential arithmetics.

	1st arithmetic (arguments)	2nd arithmetic (values)
classical calculus:	classical	classical
exponential calculus:	classical	exponential
biexponential calculus:	exponential	exponential
anaexponential calculus:	exponential	classical

The biexponential and anaexponential calculi were treated briefly in <u>Non-Newtonian</u> <u>Calculus</u>, where they were called the bigeometric and anageometric calculi, respectively. (The biexponential derivative is related to the concept called "elasticity" by economists, and the anaexponential integral turns out to be of the Stieltjes variety.)

Also of interest are the harmonic and quadratic arithmetics, which, when used in tandem with classical arithmetic, give rise to the following calculi:

	1st arithmetic (arguments)	2nd arithmetic (values)
harmonic calculus:	classical	harmonic
biharmonic calculus:	harmonic	harmonic
anaharmonic calculus:	harmonic	classical
quadratic calculus:	classical	quadratic
biquadratic calculus:	quadratic	quadratic
anaquadratic calculus:	quadratic	classical

(The harmonic and quadratic averages play important roles in the harmonic and quadratic calculi.)

The foregoing calculi were discussed briefly in <u>Non-Newtonian</u> <u>Calculus</u>.

Of course, one may use any two arithmetics, e.g., exponential arithmetic for arguments and quadratic arithmetic for values.

N O T E

1. The nonclassical arithmetics and non-Newtonian calculi should be distinguished from the nonstandard arithmetic and analysis developed by the logicians. (Nonstandard analysis is really classical calculus developed with a rigorous use of infinitesimals.)

CHAPTER 4

Graphical Interpretations

4.1 INTRODUCTION

> "Perhaps the one tendency that did
> more than any other to conceal from
> mathematicians for almost two cen-
> turies the logical basis of the
> [classical] calculus was the result
> of the attempt to make geometrical,
> rather than arithmetic, conceptions
> fundamental."
>
> Carl B. Boyer[1]

Originally we conceived the non-Newtonian calculi
analytically, not geometrically. Indeed, only at a rela-
tively late stage of our investigations were we able to
interpret the operators of the non-Newtonian calculi in
a geometric manner that we considered to be suitable.
(The suitability of such interpretations is of course a
subjective matter.) The discovery of those geometric
interpretations ultimately led to the development of non-
Cartesian geometries, one of which will be discussed in
Section 6.3.

N O T E

1. Carl B. Boyer, The History of the Calculus and Its
Conceptual Development (New York: Dover reprint, 1949),
p. 104.

4.2 EXPONENTIAL GRAPHS

By *-paper we mean paper that is ruled off in squares and labeled as follows:

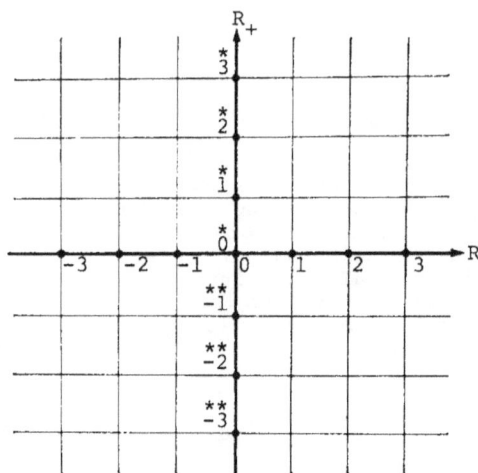

The horizontal axis is marked in the usual way, but on the vertical axis the equi-spaced dots are marked with the *-integers

$$\ldots, \overset{**}{-2}, \overset{**}{-1}, \overset{*}{0}, \overset{*}{1}, \overset{*}{2}, \ldots,$$

that is, with

$$\ldots, e^{-2}, e^{-1}, e^{0}, e^{1}, e^{2}, \ldots.$$

The origin corresponds to $(0, \overset{*}{0})$.

Although *-paper is equivalent to semi-logarithmic paper with the vertical axis labeled logarithmically, *-paper is better suited for our purposes.

The *-graph of a set of positive points is the result of plotting all its members on *-paper. Only positive points can be plotted on *-paper.

4.3 EXPONENTIAL DISTANCE

For convenience the symbol P_i will henceforth be used to denote the ordered pair of numbers (x_i, y_i).

The *-distance between positive points P_1 and P_2 is denoted by $\overset{*}{d}(P_1, P_2)$, and is defined to be the number

$$\sqrt[*]{(\overset{*}{x_1} \overset{*}{-} \overset{*}{x_2})\overset{*}{^2} \overset{*}{+} (\overset{*}{y_1} \overset{*}{-} y_2)\overset{*}{^2}} \; ,$$

which equals

$$\exp\left\{ \sqrt{(x_1 - x_2)^2 + (\ln y_1 - \ln y_2)^2} \right\} ,$$

and may be obtained graphically by plotting P_1 and P_2 on *-paper and measuring their separation with the 'ruler' provided by the upper half of the vertical axis.

It turns out that

$$\overset{*}{d}(P_1, P_2) \overset{*}{\geq} \overset{*}{0} ;$$

if $x_1 = x_2$, then $\overset{*}{d}(P_1, P_2) = |\overset{*}{y_1} \overset{*}{-} y_2|$; and

if $y_1 = y_2$, then $\overset{*}{d}(P_1, P_2) = |\overset{*}{\overset{*}{x_1}} \overset{*}{-} \overset{*}{x_2}|$.

In Section 6.3, *-distance will be discussed further.

4.4 GRAPHICAL INTERPRETATION OF EXPONENTIAL SLOPE

As expected the *-graph of each exponential function is a straight line.

Consider any exponential function $u(x) = \exp(mx + c)$ whose *-slope e^m is greater than $\overset{*}{0}$. Choose any two distinct positive points P_1 and P_2 on u, as shown in the following figure, and let P_3 be the vertex of the indicated right angle.

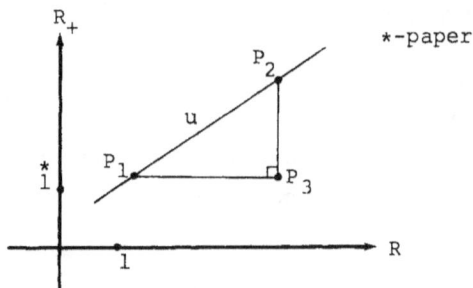

Then the ∗-slope of u equals $\overset{\star}{\text{d}}(P_2, P_3) \mathbin{\overset{\star}{/}} \overset{\star}{\text{d}}(P_1, P_3)$.

With obvious adjustments an interpretation can be given if the ∗-slope of u is less than $\overset{\star}{0}$, in which case the ∗-graph of u is decreasing. If the ∗-slope of u is $\overset{\star}{0}$, then the ∗-graph of u is horizontal.

It is not necessary to provide a separate interpretation for the ∗-gradient, since it is defined directly in terms of ∗-slope.

4.5 GRAPHICAL INTERPRETATION OF THE EXPONENTIAL DERIVATIVE

In Section 2.5 is was observed that the ∗-derivative of a function f at an argument a is equal to the ∗-slope of the exponential function u that is tangent to f at (a,f(a)). The ∗-graph would look like this:

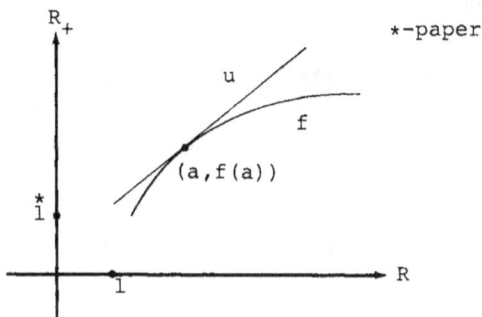

4.6 GRAPHICAL INTERPRETATION OF THE EXPONENTIAL INTEGRAL

In this section and the next all references to geometric figures are intended to apply to figures as they
appear on *-paper.

A <u>unit square</u> is a square with sides of *-length $\overset{*}{1}$.

The <u>*-area</u> of a rectangle is the *-product of its
*-length and *-width. The *-area of a unit square is $\overset{*}{1}$.
If a rectangle is decomposable into n unit squares, then its
*-area equals $\overset{*}{n}$, as expected.

Let f be a continuous function on [r,s] with values
greater than $\overset{*}{0}$. Let S be the region bounded by the *-graph
of f, the horizontal axis, and the vertical lines at r and
s.   −

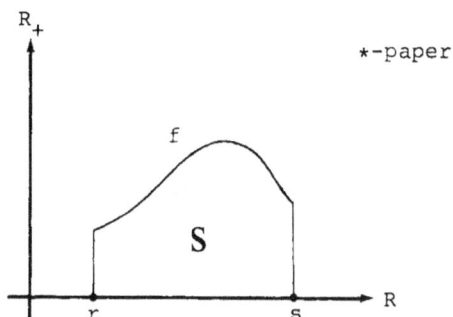

By partitioning the region S into approximating rectangles
and using the customary limit process, one can readily de-
fine the *-area of S, which turns out to be $\int_{r}^{\overset{*}{s}} f$, thereby
providing a simple interpretation of the *-integral.

4.7 GRAPHICAL INTERPRETATION OF THE GEOMETRIC AVERAGE

Let A be the *-area of the region S of the preceding section, and let m denote $\overset{*}{M}{}_r^s f$. It turns out that m is greater than $\overset{*}{0}$, and that A is equal to the *-area of the indicated rectangle.

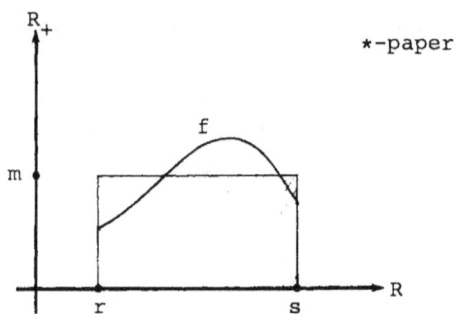

Thus, $\overset{*}{M}{}_r^s f$ is the *-height of the rectangle whose *-base is the line segment from $(r,\overset{*}{0})$ to $(s,\overset{*}{0})$ and whose *-area is equal to the *-area of S.

CHAPTER 5

Heuristic Principles of Application

5.1 Introduction

> "...all the ingenious analysis which
> has evolved from the hypothesis of
> linearity is at best a first approx-
> imation to the applicable mathematics
> of the future."
>
> E. T. Bell[1]

Since there are situations where success in applying
the classical calculus seems to depend on a felicitous
choice of simplifying assumptions, it is not unreasonable
to suppose that there are situations where the *-calculus
might prove to be useful. Accordingly, this chapter contains some heuristic principles that may be helpful for
selecting appropriate operators from the classical and *-
calculi.

Although one is always free to use any operator that
is meaningful in a given context, a suitable choice of an
operator or calculus depends chiefly upon its intended use.

N O T E

1. E. T. Bell, The Development of Mathematics, 2nd ed.
(New York: McGraw-Hill, 1945).

45

5.2 CLASSICAL AND EXPONENTIAL TRANSLATIONS

Let a and b be any two given constants. Then a
classical translation is a mapping of each point (x, y) to
the point $(x + a, y + b)$. It is convenient to conceive a
classical translation as a change of origin in the first
coordinates and a change of origin in the second coordinates.

Every classical translation of a linear function is
a linear function having the same classical slope.

Let a and b be any two given constants, with b
positive. Then a *-translation is a mapping of each posi-
tive point (x, y) to the positive point $(x + a, y \overset{*}{+} b)$, that
is, $(x + a, b\,y)$. It is convenient to conceive a *-
translation as a change of origin in the first coordinates
and a change of scale (or unit) in the second coordinates.

Every *-translation of an exponential function is an
exponential function having the same *-slope.

5.3 CHOOSING GRADIENTS AND DERIVATIVES

We shall give three heuristic principles for choosing
gradients and derivatives. The first principle requires
some preliminary discussion.

Let f be classically differentiable at a point P on
f, and let $\overline{f}$ and $\overline{P}$ be the images of f and P under any given
classical translation. Then the classical derivative of f
at P is equal to the classical derivative of $\overline{f}$ at $\overline{P}$. Briefly, the classical derivative is classically invariant. Of
course, the classical slope and gradient are also classically invariant.

Let f be a positive function that is *-differentiable
at a positive point P on f, and let $\overset{*}{f}$ and $\overset{*}{P}$ be the images of
f and P under any given *-translation. Then the *-derivative
of f at P is equal to the *-derivative of $\overset{*}{f}$ at $\overset{*}{P}$. Briefly,
the *-derivative is *-invariant. Of course, the *-slope and

*-gradient are also *-invariant.

<u>Principle I</u>

If classical (*-) invariance is desired, then
the classical (*-) gradient and derivative
should be considered.

For example, physicists naturally desire that the
average (or instantaneous) velocity of an object moving
rectilinearly be independent of the origins used for measuring time and distance, i.e., be classically invariant.
Accordingly, they define the average (or instantaneous)
velocity of the object to be the classical gradient (or
derivative) of its position function.

Likewise, it is natural to desire that the average
(or instantaneous) rate of change of the mass of a radioactive substance be independent of the origin and unit used
for measuring time and mass, respectively, i.e., be *-
invariant. Accordingly, it would be appropriate to define
that average (or instantaneous) rate of change to be the
*-gradient (or derivative) of the mass function.

It is quite possible that *-invariant gradients and
derivatives will be of considerable interest to physicists,
in view of the following remarks by John A. Wheeler:

"In the past five years one of the greatest
developments in elementary particle physics
has been so-called conformal invariance,
the discovery that the equations of elementary particle physics possess a property
that in effect is this: changing the scale
in which you examine the phenomenon does
not change the nature of the phenomenon.
For decades this so-called conformal invariance has been known to be a property of
electromagnetism. The conformal feature of
gravitation physics has been known for a
lesser time. But the conformal features of
elementary particle physics are a discovery
of the past five years. This strange feature of nature that permits us to subsume
the very large and the very small under the
same kind of equation is something we don't

yet understand, but it raises in one's mind
the perpetual question: Can it be true that
what we think of as the very small and what
we think of as the very large are really not
so different?"[1]

Principle II

Where changes in arguments and values are naturally measured by differences, the classical gradient
and derivative should probably be used. Where
changes in arguments and values are naturally measured by differences and ratios respectively, the
*-gradient and derivative are probably better
choices.

For example, the classical gradient and derivative
are appropriate for analyzing motion because changes in
time and position are measured by differences. But the *-
gradient and derivative are appropriate for analyzing stock-
price movements because changes in time and price are best
measured by differences and ratios respectively.

Principle III

Where the functional relationship between two magnitudes is, or is assumed to be, linear (exponential) under normal or ideal conditions, the
classical (*-) gradient and derivative should be
considered.

For example, the classical gradient and derivative
are useful for analyzing motion because under ideal conditions, i.e. the absence of forces, the relationship of position to time is assumed to be linear (Newton's First Law).
On the other hand, the *-gradient and derivative would be
useful for analyzing radioactive decay because the relationship of mass to time is normally exponential.

N O T E

1. Intellectual Digest, June 1973.

5.4 CHOOSING INTEGRALS

Suppose that a scientific relationship has been expressed by a differential equation involving ∗-derivatives
and perhaps operations of ∗-arithmetic. To solve that equation one would certainly try to use ∗-integration. If obtained, the solutions would initially be expressed by means
of ∗-integrals with variable upper limits, which in turn
would be evaluated in terms of known functions or new functions defined for the purpose. (Many functions were originally defined as classical integrals with variable upper
limits.) It is even conceivable that solutions in closed
form could be obtained for certain intractable differential
equations by re-expressing them with ∗-derivatives and operations of ∗-arithmetic.

A differential equation containing classical and ∗-
derivatives can readily be transformed into an equation
involving only classical derivatives or only ∗-derivatives.
(See Section 2.11.)

Integrals are also used for DEFINING sophisticated
scientific concepts, for example, the concept of work.
Suppose a point mass moves along the x-axis as a result of
a force directed along the axis. Let the force be $f(x)$ at
position x. If the force f is constant, the work on the
position interval $[r,s]$ is defined to be $(s - r) \cdot f$, which
we denote by $W_r^s f$. When f is constant the following conditions are clearly satisfied.

1. $W_r^s f = (s - r) \cdot f$.
2. Work is monotonically increasing with respect
   to force.
3. Work is additive with respect to displacement;
   that is, for any positions r, s, t such that
   $r < s < t$,
   $$W_r^s f + W_s^t f = W_r^t f .$$

Desiring to extend the concept of work to the case
where the force f is continuously variable, the physicist
stipulates that the extended work concept should satisfy
Condition 1 when f is constant and should satisfy Conditions
2 and 3 in general. The solution is now uniquely determined;
for according to Section 1.9, the operator W satisfies those
three conditions if and only if W is the classical integral.
Thus, the physicist must adopt the following definition:

$$W_r^s f = \int_r^s f \, .$$

The foregoing well-known example illustrates the fact
that the characterization of the classical integral in Section 1.9 is a heuristic principle for its appropriate use.

Now suppose that we are concerned with a positive
magnitude g, called 'gorce,' which may be continuously variable with respect to time. If g is constant, we define the
'toil' on the time interval [r,s] to be g^{s-r}, denoted by
$T_r^s g$. (Somewhere we have seen such a concept, perhaps in
economics.) When g is constant, the following conditions
are clearly satisfied.

   1. $T_r^s g = g^{s-r}$.
   2. Toil is monotonically increasing with respect
       to gorce.
   3. Toil is multiplicative with respect to time;
       that is, for any instants r, s, t such that
       r < s < t,

$$T_r^s g \cdot T_s^t g = T_r^t g \, .$$

To extend the toil concept to the case where the
gorce g is continuously variable, we stipulate that the extended toil concept should satisfy Condition 1 when g is
constant and should satisfy Conditions 2 and 3 in general.
The solution is now uniquely determined; for, according to
Section 2.9, the operator T satisfies those three conditions
if and only if T is the *-integral. Thus, the following

definition must be adopted:

$$T_r^s g = \int_r^{*s} g .$$

It should now be clear that the characterization of the *-integral in Section 2.9 is a heuristic principle for its appropriate use.

Lastly it should be mentioned that, in effect, E. C. Zeeman uses what appear to be *-integrals in his mathematical model of the brain. An elementary account of Zeeman's work is given in Mathematics in a Changing World (New York: Walker & Co., 1973) by Michael Holt and D. T. E. Marjoram, who provide bibliographic references.

5.5 Choosing Averages

Most of this section is devoted to averages of finite sequences of numbers rather than to averages of continuous functions, since a choice of the latter tacitly depends on a choice of the former. Moreover, we are not concerned here with probabilistic justifications for the use of a particular average, such as are found in the theory of errors.

Historically one reason for the popularity of the arithmetic average is its simplicity of calculation, but that issue is surely irrelevant in this age of computers. For instance, one investment advisory service has for many years maintained a geometric average of 1000 stocks.[1]

In choosing a method of averaging physical magnitudes the fundamental issue is the natural method of combining them. Where physical magnitudes are naturally combined by addition (multiplication), the arithmetic average (geometric average) is physically meaningful and possibly useful. We are not claiming, of course, that the arithmetic average is the only additive average or that the geometric average is the only multiplicative average. One should keep in mind that if magnitudes m_1 and m_2 are naturally combined by addition, then the magnitudes p^{m_1} and p^{m_2} are naturally combined by multiplication (p a positive constant).

We shall let Mu_i represent the arithmetic average of n numbers u_1, ..., u_n, and if the u_i are positive we shall let $\overset{*}{M}u_i$ represent their geometric average.

Since the geometric and arithmetic averages are the 'natural' averages in *- and classical arithmetics respectively (Section 3.3), the geometric average has the same properties relative to *-arithmetic as the arithmetic average has relative to classical arithmetic. For example, since

$$(1) \qquad M(u_i - v_i) = Mu_i - Mv_i ,$$

one should expect that

$$\overset{*}{M}(u_i \overset{*}{-} v_i) = \overset{*}{M}u_i \overset{*}{-} \overset{*}{M}v_i .$$

And indeed, the preceding equation is valid since it is merely a re-statement of the divisional character of $\overset{*}{M}$:

$$(2) \qquad \overset{*}{M}(u_i / v_i) = \overset{*}{M}u_i / \overset{*}{M}v_i .$$

Items (1) and (2) above are often cited to rationalize the heuristic principle that differences are best averaged arithmetically, but that ratios are best averaged geometrically. However, there are situations where the arithmetic average of ratios is significant.[2]

Consider the problem of estimating the area of a rectangle from n measurements x_1, ..., x_n of its length and n measurements y_1, ..., y_n of its width. The following four estimates will be considered.

$$\text{I.} \qquad E_1 = (Mx_i) \cdot (My_i)$$

$$\text{II.} \qquad E_2 = M(x_i \cdot y_i)$$

$$\text{III.} \qquad E_3 = (\overset{*}{M}x_i) \cdot (\overset{*}{M}y_i)$$

$$\text{IV.} \qquad E_4 = \overset{*}{M}(x_i \cdot y_i)$$

Because $E_1 \neq E_2$ except in isolated cases, and because $E_3 = E_4$, it would appear that here the geometric average is more appropriate than the arithmetic average. But that is to be expected since the geometric average is multiplicative and the area is the result of a multiplication. (However, a similar analysis would indicate that the arithmetic average would be more appropriate than the geometric average if one were estimating the perimeter of the rectangle.) Furthermore, Method I, which is quite popular with scientists, has another disconcerting feature: If there arose new measurements x_{n+1} and y_{n+1} such that $x_{n+1} \cdot y_{n+1} = E_1$, then Method I applied to $x_1, \ldots, x_n, x_{n+1}$ and $y_1, \ldots, y_n, y_{n+1}$ does not yield the original estimate E_1 except in trivial cases. On the other hand, if there arose new measurements x_{n+1} and y_{n+1} such that $x_{n+1} \cdot y_{n+1} = E_3 = E_4$, then Methods III and IV applied to $x_1, \ldots, x_n, x_{n+1}$ and $y_1, \ldots, y_n, y_{n+1}$ yield the original estimate $E_3 (= E_4)$.

Now consider the problem of estimating the density of an object, given n measurements of its mass $w_1, \ldots, w_n$ and n measurements of its volume $v_1, \ldots, v_n$. There are at least four estimates of the density worthy of consideration.

$$\mathcal{E}_1 = Mw_i \ / \ Mv_i$$
$$\mathcal{E}_2 = M(w_i \ / \ v_i)$$
$$\mathcal{E}_3 = \overset{*}{M}w_i \ / \ \overset{*}{M}v_i$$
$$\mathcal{E}_4 = \overset{*}{M}(w_i \ / \ v_i)$$

Since $\mathcal{E}_1 \neq \mathcal{E}_2$ except in isolated cases, and since $\mathcal{E}_3 = \mathcal{E}_4$ in general, it seems that the choice ought to be the geometric average.

Scientists often invoke a 'normalcy' principle for making an appropriate choice of averages. For instance, it is often stated that the geometric average is appropriate where the functional relationship would be exponential under normal conditions. Presumably that idea arose from the

observation that if u is an exponential function, then the
value of u at the midpoint of [r,s] is equal to the geometric average of u(r) and u(s). By the same token, one would
choose the arithmetic average where the functional relationship would normally be linear, for if v is a linear function, then the value of v at the midpoint of [r,s] is equal
to the arithmetic average of v(r) and v(s).

Some scientists, notably the psychophysicist S. S.
Stevens of Harvard University, favor the use of certain invariance principles for choosing averages.[3]

As in the case for integrals, heuristic principles
for the use of averages are also provided by their characterizations (Sections 1.6 and 2.6).

In many situations, averages are intuitively more
satisfying than integrals. Consider, for instance, a particle moving rectilinearly with positive velocity v at time
t. The distance s traveled in the time interval [a,b] is
given by

$$s = \int_a^b v \,.$$

Although that fact may be clear to a student, he may nevertheless find that the following formula conveys a more
immediate meaning:

$$s = (b - a) \cdot M_a^b v \,;$$

that is, the distance traveled equals the product of the
time elapsed and the arithmetic average of the velocity.
This version is a direct extension of the case where v is
constant. Other examples will readily occur to the reader.
(For the 'toil' concept in the preceding section, we may
write

$$T_r^s g = [\overset{*}{M}_r^s g]^{s-r} \,,$$

a formula that is a direct extension of the case where g is constant.)

N O T E S

1. American Investors Service (Greenwich, Connecticut) distribute an interesting booklet by George A. Chestnut, Jr., who gives some excellent reasons why he considers geometric averaging the better method of averaging stock prices.

2. For example, suppose that initially $1000 is invested in one stock at $10 per share and $1000 in another stock at $20 per share. Subsequently the stocks are worth $5 and $50 per share respectively. Since the original investment of $2000 increased in value to $3000, the overall ratio change in value is 1.5, which equals the ARITHMETIC average of the ratio changes, 0.5 and 2.5, for the individual stocks.

3. A detailed discussion is given by Stevens in his article "On the Averaging of Data," which appeared in Science, Vol. 121 (January 28, 1955), pp. 113-6. Some comments on Stevens' ideas may be found in Brian Ellis' book, Basic Concepts of Measurement (Cambridge University Press, 1966).

5.6 Constants And Scientific Concepts

> "Where there is change, he [the scientist] looks for constancy in the rate of change; and failing that, for constancy in the rate of the rate of change."
>
> Nelson Goodman[1]

In this section we illustrate the thesis that the invention of a scientific concept may depend on the isolation or discovery of a suitable constant; and we suggest that new scientific concepts may arise from the constants provided by the *-slopes of exponential functions.

Consider the concept of average speed. The definition "distance traveled per unit time" is incomplete because it fails to provide a method of determining the average speed of an accelerated particle. The definition "distance divided by time," though not incorrect, is a gross oversim-

plification that fails to reveal the underlying issues.
Fortunately there is a completely satisfactory definition,
which was undoubtedly known to Galileo.

Let us begin with Galileo's definition of uniform
motion as "one in which the distances traversed...during
ANY equal intervals of time are themselves equal." (Continuity of the motion is tacitly intended here.)[2] Then we
isolate a constant in each given uniform motion by defining
speed to be the distance traveled in any unit time-interval.
Finally, for a particle that moves non-uniformly a distance
d in time t, we define the average speed to be the speed
that a particle in uniform motion must have in order to
travel a distance d in time t. In our opinion, neither the
simplicity nor the obviousness of the answer, d/t , justi-
fies its use as the definition of average speed.

Although Archimedes undoubtedly knew, in effect, the
definition of speed in uniform motion, it was probably
Galileo who first introduced the concepts of speed and
acceleration in non-uniform motion, and he may have been
the first to define uniform motion rigorously.

The critical first step in defining average speed is
the isolation of the constant (speed) in the phenomenon of
uniform motion. Similarly, the critical step in defining
the *-gradient is the identification of the constant (*-
slope) in an exponential function.

Wherever a scientific phenomenon, or an idealized
version thereof, is describable by an exponential function,
one automatically has a fundamental constant, the *-slope,
which may prove to be useful.

For example, in the idealized model of radioactive
decay, the relationship of mass to time is describable by
an exponential function whose *-slope provides a constant
that is independent of the unit of mass and equals 1 more
than the percentage change of the mass on any unit time-
interval.

As another example, consider the interesting, but still disputed, law of Edwin Hubble:

> Distant galaxies are receding
> at speeds in proportion to
> their distance from us.

From that law, it follows that the relationship of distance (between a galaxy and us) to time is describable by an exponential function. The *-slope of that exponential function, which might aptly be called *-speed, provides a constant that is independent of the unit of distance and equals 1 more than the percentage change of the distance on any unit time-interval. Indeed, Hubble's Law may be simply restated as follows:

> Distant galaxies are receding
> at constant *-speeds.

N O T E S

1. Nelson Goodman, <u>Problems</u> <u>and</u> <u>Projects</u> (Bobbs-Merrill, 1972), p. 351.

2. This is the first formal definition in Galileo's masterpiece <u>Two</u> <u>New</u> <u>Sciences</u>, "Third Day," translated by Henry Crew and Alfonso de Salvio. We have capitalized the word "any" for emphasis, particularly since Galileo adds a Caution on this very point. Galileo uses a similar technique to define uniformly accelerated motion.

CHAPTER 6

Exponential Geometry: A Non-Cartesian System

6.1 INTRODUCTION

> "In itself, no curve is simpler
> than another."
>
> Charles S. Peirce[1]

It is well-known that for Euclidean plane geometry one can construct an arithmetic model in which every linear function represents a Euclidean line. The chief purpose of this chapter is to indicate how one can also construct an arithmetic model for Euclidean plane geometry in which every exponential function represents a Euclidean line.

The construction of arithmetic models for geometry is of more than purely mathematical interest, as the following remark by the physicist John L. Synge suggests:

> "The relationship of geometry to relativity is
> most satisfactorily established when geometry
> is regarded analytically, a 'point' being
> nothing but a [sequence] of numbers... and a
> 'line' a set of points. This is a most fruit-
> ful way to look at geometry...."[2]

Since we are concerned here solely with Euclidean geometry of the plane, adjectives such as "plane" and "two-dimensional" will be omitted. Nevertheless, our development can be extended to n-dimensions.

The symbol P_i will still be used to represent the point (x_i, y_i).

N O T E S

1. The Monist (January 1891).

2. John L. Synge, <u>Relativity</u>: The <u>Special</u> <u>Theory</u>, 2nd ed. (North-Holland Publishing Co., 1964), p. 4.

6.2 CARTESIAN GEOMETRY

In his classic <u>Grundlagen</u> <u>der</u> <u>Geometrie</u> (1899), Hilbert constructed for Euclidean geometry an arithmetic model that he called "the ordinary analytic Cartesian geometry," here abbreviated to "Cartesian geometry." Though Hilbert's construction is elegant, a different approach will be used in this section.[1]

<u>Cartesian</u> <u>geometry</u> is the system consisting of the set of all points (i.e. ordered pairs of numbers) in which the <u>classical</u> <u>distance</u> $d(P_1, P_2)$ between points P_1 and P_2 is stipulated to be the nonnegative number

$$\sqrt{(x_1 - x_2)^2 + (y_1 - y_2)^2} \ .$$

Within Cartesian geometry one can define counterparts of all Euclidean notions, some of which will be discussed in this section.

Classical distance is classically invariant; that is, the classical distance between points is unchanged under all classical translations.

The classical-distance operator d is a metric because of the following result.

<u>Theorem</u>

> For all points P_1, P_2, and P_3,
> $d(P_1, P_2) \geqq 0$,
> $d(P_1, P_2) = 0$ if and only if $P_1 = P_2$,
> $d(P_1, P_2) = d(P_2, P_1)$,
> $d(P_1, P_2) + d(P_2, P_3) \geqq d(P_1, P_3)$.

The points P_1, P_2, and P_3 are <u>collinear</u> provided that at least one of the following holds:

$$d(P_2, P_1) + d(P_1, P_3) = d(P_2, P_3),$$
$$d(P_1, P_2) + d(P_2, P_3) = d(P_1, P_3),$$
$$d(P_1, P_3) + d(P_3, P_2) = d(P_1, P_2).$$

A <u>line</u> is a set L of at least two distinct points such that for all distinct points P_1 and P_2 in L, a point P_3 is in L if and only if P_1, P_2, and P_3 are collinear.

A set of points is <u>vertical</u> provided that all its members have the same first coordinate.

<u>Theorem</u>

        The class of nonvertical lines is identical with the class of linear functions.

In view of the preceding theorem one may say that Cartesian geometry is a linear model of Euclidean geometry.

Two lines are <u>parallel</u> provided they are identical or have no common point.

<u>Theorem</u>

        Two nonvertical lines are parallel if and only if they have the same classical slope.

A line L_1 is <u>perpendicular</u> to a line L_2 provided they have a common point P, and for any points P_1 on L_1 and P_2 on L_2, distinct from P,

$$d(P_1, P) < d(P_1, P_2).$$

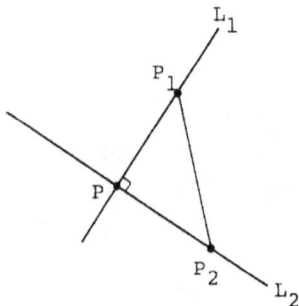

Theorem

> If two perpendicular lines L_1 and L_2 intersect at point P, then for all points P_1 on L_1 and P_2 on L_2,
>
> $$[d(P_1, P)]^2 + [d(P, P_2)]^2 = [d(P_1, P_2)]^2.$$

Theorem

> Two nonvertical lines are perpendicular if and only if the product of their classical slopes is -1.

N O T E

1. Hilbert's construction is neatly summarized in Howard Eves' book, A Survey of Geometry, Vol.II (Allyn and Bacon, 1965), pp. 102-5. Hilbert's objective was to prove that Euclidean geometry is consistent with classical arithmetic, but that is not our concern here.

6.3 EXPONENTIAL GEOMETRY

> "...the distance concept is logically arbitrary."
>
> Einstein

By *-geometry we mean the system consisting of the set of all positive points (i.e. points with positive second coordinates) in which the *-distance $\overset{*}{d}(P_1, P_2)$ between two positive points P_1 and P_2 is stipulated to be the number

$$\overset{*}{\sqrt{(x_1 \overset{*}{-} x_2)^{\overset{*}{2}} \overset{*}{+} (y_1 \overset{*}{-} y_2)^{\overset{*}{2}}}} ,$$

which equals

$$\exp\left\{ \sqrt{(x_1 - x_2)^2 + (\ln y_1 - \ln y_2)^2} \right\} .$$

(*-distance was introduced and interpreted graphically in

Section 4.3.)

Within *-geometry one can define counterparts of all Euclidean notions, some of which will be discussed in this section.

*-distance is *-invariant; that is, the *-distance between positive points is unchanged under all *-translations.

Although $\overset{*}{d}$ is not a metric in the usual sense, it is a *-metric in the sense indicated by the following.

Theorem

For all positive points P_1, P_2, and P_3,

$$\overset{*}{d}(P_1, P_2) \geqq \overset{*}{0},$$

$$\overset{*}{d}(P_1, P_2) = \overset{*}{0} \text{ if and only if } P_1 = P_2,$$

$$\overset{*}{d}(P_1, P_2) = \overset{*}{d}(P_2, P_1),$$

$$\overset{*}{d}(P_1, P_2) \overset{*}{+} \overset{*}{d}(P_2, P_3) \geqq \overset{*}{d}(P_1, P_3).$$

The positive points P_1, P_2, and P_3 are *-collinear provided that at least one of the following holds:

$$\overset{*}{d}(P_2, P_1) \overset{*}{+} \overset{*}{d}(P_1, P_3) = \overset{*}{d}(P_2, P_3),$$

$$\overset{*}{d}(P_1, P_2) \overset{*}{+} \overset{*}{d}(P_2, P_3) = \overset{*}{d}(P_1, P_3),$$

$$\overset{*}{d}(P_1, P_3) \overset{*}{+} \overset{*}{d}(P_3, P_2) = \overset{*}{d}(P_1, P_2).$$

If $x_1 = x_2 = x_3$ or $y_1 = y_2 = y_3$, then P_1, P_2, and P_3 are *-collinear, but other circumstances also engender *-collinearity, as will be seen shortly.

A *-line is a set L of at least two distinct positive points such that for any distinct positive points P_1 and P_2 in L, a positive point P_3 is in L if and only if P_1, P_2, and P_3 are *-collinear.

Theorem

The class of nonvertical *-lines is identical with the class of exponential functions.[1]

In view of the preceding theorem, one may say that *-geometry is a nonlinear model of Euclidean geometry. (Infinitely-many nonlinear models of Euclidean geometry are constructed in Chapter 10 of <u>Non-Newtonian Calculus</u>.)

Two *-lines are *-parallel provided they are identical or have no common point. For example, the *-lines with equations $y = \exp(x + 1)$ and $y = \exp(x + 2)$ are *-parallel.

Theorem

Two nonvertical *-lines are *-parallel if and only if they have the same *-slope.

Of course, a *-line that is *-parallel to the horizontal axis (on *-paper) has *-slope equal to $\overset{*}{0}$.

A *-line L_1 is *-perpendicular to a *-line L_2 provided they have a common positive point P, and for any positive points P_1 on L_1 and P_2 on L_2, distinct from P,

$$\overset{*}{d}(P_1, P) < \overset{*}{d}(P_1, P_2).$$

Theorem

If two *-perpendicular *-lines L_1 and L_2 intersect at positive point P, then for any positive points P_1 on L_1 and P_2 on L_2,

$$[\overset{*}{d}(P_1, P)]^{\overset{*}{2}} \overset{*}{+} [\overset{*}{d}(P, P_2)]^{\overset{*}{2}} = [\overset{*}{d}(P_1, P_2)]^{\overset{*}{2}}.$$

Theorem

Two nonvertical *-lines are *-perpendicular if and only if the *-product of their *-slopes is $\overset{**}{-1}$.

The *-arc-length of a suitably behaved arc can be defined with the usual limit process by partitioning the arc and using the *-sum of the *-distances between successive positive points. If a positive function f has a continuous *-derivative on [r,s], then the *-arc-length of f on [r,s] turns out to be

$$\int_{r}^{*s} \, ^* \sqrt{\overset{*}{1} \overset{*}{+} (\overset{*}{D}f)^{\overset{*}{2}}} \;\; .$$

As expected, the *-distance between two positive points P_1 and P_2 is equal to the *-arc-length of the *-line segment connecting P_1 and P_2.

Though the Cartesian and *-geometries are structurally equivalent, it is just as profitable to distinguish them as it is to distinguish the classical and *-calculi.

N O T E

1. The definition of "vertical" given in Section 6.2 applies equally well here.

CHAPTER 7

Exponential Vectors and Centroids

7.1 Exponential Vectors

The set of all classical translations can be made into a vector space in the following way. For any numbers a and b, let the classical translation that maps each point (x,y) to (x + a, y + b) be denoted by v[a;b] and be called a <u>classical</u> <u>vector</u>. Define the <u>sum</u> of v[a;b] and v[c;d] to be v[a + c; b + d], and define the <u>scalar</u> <u>product</u> of a number k and v[a;b] to be v[ka; kb]. Then this system is a vector space over the field of classical arithmetic. One may also define the <u>norm</u> of v[a;b] to be the number $\sqrt{a^2 + b^2}$, which equals the classical distance between (a,b) and (0,0).

Every classical vector v[a;b] is <u>rectilinear</u> in the sense that for any point (x,y), the following points are collinear: (x,y), its image (x + a, y + b), and the latter's image (x + a + a, y + b + b).

The set of all *-translations can also be made into a vector space. For any number a and positive number b, let the *-translation that maps each positive point (x,y) to (x + a, y $\overset{*}{+}$ b) be denoted by $\overset{*}{v}$[a;b] and be called a *-<u>vector</u>. Define the <u>sum</u> of $\overset{*}{v}$[a;b] and $\overset{*}{v}$[c;d] to be $\overset{*}{v}$[a + c; b $\overset{*}{+}$ d], and define the <u>scalar</u> <u>product</u> of a positive number k and $\overset{*}{v}$[a;b] to be $\overset{*}{v}$[(ln k)a; k $\overset{*}{\times}$ b]. Then this system is a vector space over the field of *-arithmetic. One may also define the <u>norm</u> of $\overset{*}{v}$[a;b] to be the number $\overset{*}{\sqrt{}}\left(\overset{*}{a}\right)^{\overset{*}{2}} \overset{*}{+} b^{\overset{*}{2}}$, which equals the *-distance between (a,b) and (0,$\overset{*}{0}$).

Each *-vector $\overset{*}{v}[a;b]$ is *-rectilinear in the sense
that for any positive point (x,y), the following positive
points are *-collinear: (x,y), $(x + a, y \overset{*}{+} b)$, and
$(x + a + a, y \overset{*}{+} b \overset{*}{+} b)$. If $a = 0$ or $b = 0$, then $\overset{*}{v}[a;b]$ is
rectilinear because for any positive point (x,y), the posi-
tive points (x,y), $(x + a, y \overset{*}{+} b)$, and $(x + a + a, y \overset{*}{+} b \overset{*}{+} b)$
are collinear. However, if $a \neq 0$ and $b \neq 0$, then $\overset{*}{v}[a;b]$ is
curvilinear because it is not rectilinear. Thus, most *-
vectors are curvilinear.

7.2 EXPONENTIAL CENTROIDS

In a document sent to Eratosthenes, Archimedes explained in detail how he used classical centroids to discov-
er deep geometric theorems.[1] Believing that centroids will
one day again prove to have considerable heuristic value in
pure mathematics, we devote this section to a novel definition of the classical centroid and to a definition of the
new *-centroid. Both centroids will be used in Chapter 8
on least-squares methods.

Since the classical distance from a point $P(x,y)$ to
a point $P_1(x_1, y_1)$ is

$$\sqrt{(x - x_1)^2 + (y - y_1)^2} ,$$

it is not unreasonable to define the classical distance
from a point $P(x,y)$ to n points $P_1(x_1, y_1), \ldots, P_n(x_n, y_n)$
to be

$$\sqrt{[(x - x_1)^2 + (y - y_1)^2] + \cdots + [(x - x_n)^2 + (y - y_n)^2]} ,$$

or simply

$$\sqrt{[d(P, P_1)]^2 + \cdots + [d(P, P_n)]^2} .$$

For n = 1, the preceding expression reduces to $d(P, P_1)$, as it should.

The idea for the following theorem was suggested by an article published by Legendre in 1805.[2]

Theorem

> For any n points $P_1, \ldots, P_n$, there is a unique point C that is classically closer than every other point to $P_1, \ldots, P_n$, and the coordinates of C are $(x_1 + \cdots + x_n)/n$ and $(y_1 + \cdots + y_n)/n$.

The point C in the preceding theorem is called the classical centroid of the points $P_1, \ldots, P_n$, and has the following important property:

> (1) For each linear function u containing C,
> $$[u(x_1) - y_1] + \cdots + [u(x_n) - y_n] = 0.$$

It is an interesting fact, which was known in substance by Archimedes, that the classical centroid of any given n points of a linear function is contained in that function.

If C is the classical centroid of $P_1, \ldots, P_n$, and $\overline{C}, \overline{P}_1, \ldots, \overline{P}_n$ are their images under any given classical translation, then $\overline{C}$ is the classical centroid of $\overline{P}_1, \ldots, \overline{P}_n$.

Our development of the *-centroid is similar to that of the classical centroid.

The *-distance from a positive point P(x,y) to n positive points $P_1(x_1, y_1), \ldots, P_n(x_n, y_n)$ is defined to be

$$\sqrt[*]{[\overset{*}{d}(P, P_1)]^{\overset{*}{2}} \overset{*}{+} \cdots \overset{*}{+} [\overset{*}{d}(P, P_n)]^{\overset{*}{2}}} \,,$$

which reduces to $\overset{*}{d}(P, P_1)$ for n = 1.

Theorem

> For any n positive points $P_1, \ldots, P_n$, there is a unique positive point C that is *-closer than every other positive point to $P_1, \ldots, P_n$, and

the coordinates of C are $(x_1 + \cdots + x_n) / n$ and $(y_1 \overset{*}{+} \cdots \overset{*}{+} y_n) \overset{*}{/} \overset{*}{n}$. (The second coordinate of C is, of course, the geometric average of the y_i.)

The positive point C in the preceding theorem is the *-centroid of the positive points $P_1, \ldots, P_n$, and has the following property:

(2) For each exponential function u containing C,
$$[u(x_1) \overset{*}{-} y_1] \overset{*}{+} \cdots \overset{*}{+} [u(x_n) \overset{*}{-} y_n] = \overset{*}{0}.$$

The *-centroid of any given n positive points of an exponential function is contained in that function.

If C is the *-centroid of n positive points $P_1, \ldots, P_n$, and $\overset{*}{C}, \overset{*}{P_1}, \ldots, \overset{*}{P_n}$ are their images under any given *-translation, then $\overset{*}{C}$ is the *-centroid of $\overset{*}{P_1}, \ldots, \overset{*}{P_n}$.

N O T E S

1. A palimpsest of Archimedes' document, commonly called The Method, was discovered in 1906 by J. L. Heiberg.

2. Legendre's article is entitled "Sur la Méthode des moindres quarrés," a translation of which appears in A Source Book in Mathematics, Vol. II, ed. David E. Smith (Dover reprint, 1959), pp. 576-9.

CHAPTER 8

The Exponential Method of Least Squares

8.1 INTRODUCTION

> [The classical method of least
> squares is] "one of the most
> fundamental cornerstones of
> statistical methods."
>
> Cornelius Lanczoz[1]

Ever since the classical method of least squares was
developed by Gauss, Laplace, Legendre, and others, there have
been disagreements over the method of measuring the deviation
between an estimate and the 'true' value. Laplace claimed
that the absolute value of the difference should be used.
Gauss maintained that the square of the difference is a better choice and he made the following remark in 1820:

> "If you object that this is arbitrary, we
> readily agree. The question with which we
> are concerned is vague in its very nature;
> it can only be made precise by pretty arbi-
> trary principles. Determining a magnitude
> by observation can justly be compared to a
> game in which there is a danger of loss but
> no hope of gain.... But if we do so, to
> what can we compare an error which has actu-
> ally been made? That question is not clear,
> and its answer depends in part on conven-
> tion. Evidently the loss in the game can't
> be compared directly to the error which has
> been committed, for then a positive error
> would represent a loss, and a negative er-
> ror a gain. The magnitude of the loss must
> on the contrary be evaluated by a function
> of the [difference], whose value is always
> positive. Among the infinite number of func-
> tions satisfying these conditions, it seems
> natural to choose the simplest, which is,
> beyond contradiction, the square of the
> [difference]."[2]

Despite the many sophisticated attempts to justify Gauss'
choice by various statistical-probabilistic techniques, some
mathematicians still contend that the least-squares criteri-
on is purely arbitrary. Although we are not concerned here
with this controversy, we shall attempt to present the classical method of least squares in a manner that seems to be
natural. And then we shall use *-arithmetic to explain the
*-method of least squares, which provides a rationale for a
technique well-known to scientists.

N O T E S

1. Cornelius Lanczoz, Albert Einstein and the Cosmic World
Order (Interscience, 1965), p. 65.
2. Gauss' remark is quoted in Ian Hacking's book, Logic of
Statistical Inference (Cambridge University Press, 1965),
p. 175.

8.2 THE CLASSICAL METHOD OF LEAST SQUARES

> "Begin with the simplest examples."
>
> Hilbert[1]

Let f be a discrete function with arguments a_1,
..., a_n, and let h be any function whose arguments include
all the a_i. Since h is to be conceived as an 'estimate'
for f, our first problem is to settle on a suitable method
of defining the 'distance' or 'deviation' between h and f
in terms of their values at the a_i. Therefore we turn our
attention to the sequences

$$h(a_1), \ldots, h(a_n)$$

and

$$f(a_1), \ldots, f(a_n).$$

Each of those two sequences of n numbers may be conceived
as a point in n-dimensional Euclidean space, in which

the distance between the two points is given by

$$\sqrt{[h(a_1) - f(a_1)]^2 + \cdots + [h(a_n) - f(a_n)]^2} \; .$$

Accordingly we choose that number to represent the classical distance between the given functions h and f. For n = 1, the preceding expression reduces to the usual vertical distance $|h(a_1) - f(a_1)|$ between the points $(a_1, h(a_1))$ and $(a_1, f(a_1))$.

Now we are prepared to state the theorem that will motivate an important definition.

Theorem

For each discrete function f with at least two distinct points, there is a unique linear function u that is classically closer than every other linear function to f.

The linear function u in the foregoing theorem is called the linear function that is best-fitted to f by the classical method of least squares, henceforth abbreviated as the linear function that best-fits f classically; and u has two important properties, which, however, do not characterize it:[2]

(1) $[u(a_1) - f(a_1)] + \cdots + [u(a_n) - f(a_n)] = 0.$
(2) u contains the classical centroid of f.

Thus, the linear function that best-fits f classically is the linear function classically closest to f; and that linear function contains the classical centroid of f, which is the point classically closest to f.

The classical slope of the linear function that best-fits f classically is often used to indicate the "overall linear direction" of f.

The foregoing presentation is a natural extension of the idea that motivated our definition of the classical centroid in Section 7.2.

Now let us consider the problem of classically fitting nonlinear functions to a discrete function f with arguments $a_1, \ldots, a_n$. Consider any set S of functions whose arguments include all the a_i. If there exists a unique member h of S that is classically closer than every other member of S to f, then h is called, for brevity, the member of S that best-fits f classically. (Please note that the members of S are not assumed to be linear or discrete.)

For example, scientists often wish to fit an exponential function to f, assuming f is positive in this case. (Here S is the set of all exponential functions.) Although the classical method of least squares may be used to solve this problem, the calculations are usually quite messy. Accordingly, scientists often adopt what seems to be a 'makeshift' technique of using exp(u), where u is the linear function that best-fits ln f classically. In Section 8.4 we shall return to that 'makeshift' technique.

N O T E S

1. In her biography Hilbert (Springer-Verlag, 1970), Constance Reid reports that Hilbert said this to Hermann Weyl at the beginning of a mathematical discussion.

2. The reader may wish to compare item (1) with the items labeled (1) in Sections 3.3 and 7.2.

8.3 THE EXPONENTIAL METHOD OF LEAST SQUARES

Let f be a discrete positive function with arguments $a_1, \ldots, a_n$.

If h is any positive function whose arguments include all the a_i, then the *-distance between h and f is defined to be

$$\sqrt[*]{[h(a_1) \overset{*}{-} f(a_1)]^{\overset{*}{2}} \overset{*}{+} \cdots \overset{*}{+} [h(a_n) \overset{*}{-} f(a_n)]^{\overset{*}{2}}} \; .$$

For $n = 1$, the preceding expression reduces to the $*$-distance $|h(a_1) \overset{*}{-} f(a_1)|$ between $(a_1, h(a_1))$ and $(a_1, f(a_1))$.

Theorem

> For each discrete positive function f with at least two distinct points, there is a unique exponential function u that is $*$-closer than every other exponential function to f.

The exponential function u in the foregoing theorem is called the exponential function that is best-fitted to f by the $*$-method of least squares, abbreviated as the exponential function that best-fits f exponentially; and u has two important properties, which, however, do not characterize it:[1]   —

(1) $[u(a_1) \overset{*}{-} f(a_1)] \overset{*}{+} \cdots \overset{*}{+} [u(a_n) \overset{*}{-} f(a_n)] = \overset{*}{0}.$

(2)   u contains the $*$-centroid of f.

Thus, the exponential function that best-fits f exponentially is the exponential function $*$-closest to f; and that exponential function contains the $*$-centroid of f, which is the positive point $*$-closest to f.

The $*$-slope of the exponential function that best-fits f exponentially may be useful for indicating the "overall $*$-direction" of f.

Now let S be any set of positive functions whose arguments include all the a_i. If there exists a unique member h of S that is $*$-closer than every other member of S to f, then h will be called, for brevity, the member of S that best-fits f exponentially. (Please note that the members of S are not assumed to be discrete or exponential.)

The $*$-method of least squares is a natural extension of the idea that motivated our definition of the $*$-centroid in Section 7.2.

N O T E

1. The reader may wish to compare item (1) with the items labeled (3) in Section 3.3 and (2) in Section 7.2.

8.4 THE RELATIONSHIP BETWEEN THE TWO METHODS

Let f be a discrete positive function, and let S be any set of positive functions whose arguments include all the arguments of f. Let $\overline{f}$ = ln f, and let $\overline{S}$ consist of the natural logarithms of all the functions in S. Let $\overset{*}{L}(S;f)$ be the member of S that best-fits f exponentially, if it exists; and let $L(\overline{S};\overline{f})$ be the member of $\overline{S}$ that best-fits $\overline{f}$ classically, if it exists.

Then $\overset{*}{L}(S;f)$ and $L(\overline{S};\overline{f})$ coexist, and if they do exist,

$$\overset{*}{L}(S;f) = \exp\left\{ L(\overline{S};\overline{f}) \right\} ,$$

a relationship whose pattern is already familiar from Section 2.11.

Thus, if $\overset{*}{L}(S;f)$ exists, it can be found by using the classical method to select the member of $\overline{S}$ that best-fits $\overline{f}$ and then by applying exp to the result. Accordingly, the 'makeshift' technique described in Section 8.2 is but the *-method for fitting an exponential function to a given discrete positive function, a fact that astonished us when we first noticed it.

CHAPTER 9

Collateral Issues

9.1 INTRODUCTION

This last chapter contains, among other things, brief discussions of the percentage derivative and *-complex numbers. The reader might enjoy investigating other matters related to the *-calculus; for example, *-differential equations, Taylor series in the context of *-calculus, *-calculus of functions of several real variables, the theory of functions of a *-complex variable, or the application of *-calculus to science and engineering.

9.2 THE PERCENTAGE DERIVATIVE

The percentage derivative is closely related to the *-derivative and stems from the idea of comparing function values by percentage changes.

Each exponential function u has the following property:

For any intervals $[r_1, s_1]$ and $[r_2, s_2]$, if $s_1 - r_1 = s_2 - r_2$, then

$$\frac{u(s_1) - u(r_1)}{u(r_1)} = \frac{u(s_2) - u(r_2)}{u(r_2)} ;$$

that is, equal differences in arguments yield equal percentage changes in values.

In particular, the number $[u(b) - u(a)]/u(a)$ is the same for any numbers a and b such that $b - a = 1$, a fact which suggests the following definition.

The percentage slope of an exponential function u is the number $[u(b) - u(a)]/u(a)$, where a and b are any two numbers such that $b - a = 1$.

Now let f be a positive function. The percentage gradient of f on [r,s] is the percentage slope of the exponential function containing $(r, f(r))$ and $(s, f(s))$, and turns out to be

$$\left[\frac{f(s)}{f(r)} \right]^{\frac{1}{s-r}} - 1 \, ,$$

which equals $(\overset{*s}{G_r}f) - 1$ and is greater than -1.

In securities analysis and other applications the quantity $(\overset{*s}{G_r}f) - 1$ is often referred to as a compound growth rate. For example, suppose that one paid $64 for a share of stock. Three years later the price per share was $216. The compound (annual) growth rate of the price on the time-interval [0,3] is given by

$$\left[\frac{216}{64} \right]^{\frac{1}{3-0}} - 1 \, ,$$

which equals 50%. The significance of that figure stems from the fact that if an original investment of $64 increased 50% in each of three successive years, the final value would be $216.

The percentage derivative of f at an argument a is the following limit, if it exists and is greater than -1:

$$\lim_{x \to a} \left\{ \left[\frac{f(x)}{f(a)} \right]^{\frac{1}{x-a}} - 1 \right\} \, .$$

Clearly the percentage derivative of f at a equals 1 less than the *-derivative of f at a.

9.3 EXPONENTIAL COMPLEX-NUMBERS

The *-complex-number system consists of all positive points, for which the following two operations are defined:

$$(a_1, b_1) \oplus (a_2, b_2) = (a_1 + a_2, b_1 \overset{*}{+} b_2),$$

$$(a_1, b_1) \otimes (a_2, b_2) =$$
$$\left(\bar{a}_1 a_2 - (\ln b_1)(\ln b_2), \quad (\overset{*}{a}_1 \overset{*}{\times} b_2) \overset{*}{+} (\overset{*}{a}_2 \overset{*}{\times} b_1)\right).$$

It is interesting to note that

$$(a_1, b_1) \oplus (a_2, b_2) = (a_1 + a_2, b_1 \cdot b_2),$$

$$(a_1, b_1) \otimes (a_2, b_2) =$$
$$\left(a_1 a_2 - (\ln b_1)(\ln b_2), \quad b_2^{a_1} \cdot b_1^{a_2}\right).$$

Because the system is a field, one can define subtraction and division, which are 'inverses' of $\oplus$ and $\otimes$.

Since
$$(0, \overset{*}{1}) \otimes (0, \overset{*}{1}) = (-1, \overset{*}{0}),$$

we define $\overset{*}{i}$ to be $(0, \overset{*}{1})$. Of course, it is also true that

$$(0, \overset{*}{\underset{-}{1}}) \otimes (0, \overset{*}{\underset{-}{1}}) = (-1, \overset{*}{0}).$$

The *-modulus of a positive point (a,b) is defined to be the *-distance between (a,b) and $(0, \overset{*}{0})$, which equals

$$\overset{*}{\sqrt{(\overset{*}{a})^{\overset{*}{2}} \overset{*}{+} b^{\overset{*}{2}}}}.$$

9.4 An Insight By Boscovich

J. F. Scott of St. Mary's College, London, has observed that the following astonishing quotation indicates Roger
Joseph Boscovich (1711 - 1787) was capable of grasping the
possibility of a non-Euclidean geometry.

"But if some mind very different from ours
were to look upon some property of some curved
line as we do on the evenness of a straight
line, he would not recognize as such the evenness of a straight line; nor would he arrange
the elements of his geometry according to that
very different system, and would investigate
quite other relationships as I have suggested
in my notes.
"We fashion our geometry on the properties
of a straight line because that seems to us to
be the simplest of all. But really all lines
that are continuous and of a uniform nature are
just as simple as one another. Another kind
of mind which might form an equally clear men-
tal perception of some property of any one of
these curves, as we do of the congruence of a
straight line, might believe these curves to
be the simplest of all, and from that property
of these curves build up the elements of a
very different geometry, referring all other
curves to that one, just as we compare them to
a straight line. Indeed, these minds, if they
noticed and formed an extremely clear perception of some property of, say, the parabola,
would not seek, as our geometers do, to rectify
the parabola, they would endeavour, if one may
coin the expression, to parabolify the straight
line."

This quotation indicates to us that Boscovich was
capable of grasping the possibility of a non-Newtonian
calculus.[1]

N O T E

1. In April 1973 Robert Katz and I encountered the Boscovich quotation and Scott's remark in his article "Boscovich's
Mathematics," which appears in Lancelot Law Whyte's anthology Roger Joseph Boscovich (George Allen & Unwin Ltd., and

Fordham University Press, 1961). Scott apparently obtained
the Boscovich quotation from H. V. Gill's book, Roger Joseph
Boscovich: Forerunner of Modern Physical Theories (Dublin,
1941), pp. 50-53.
Perhaps this disclosure of Boscovich's insight will in-
spire some historian to investigate the matter fully, especially since Boscovich refers to his 'notes' on the subject.
The article on Boscovich in the magnificent Dictionary of
Scientific Biography, edited by Charles Coulston Gillispie,
sheds little light on the issue at hand.

9.5 Conclusion

> "In this century especially, mathe-
> maticians have tended to focus on
> very general classes of systems,
> and the theorems concern properties
> that are true of all or of large
> subclasses of them; however, these
> results do not usually provide much
> detailed information about any par-
> ticular member of the class."
>
> Luce & Suppes[1]

Although a general theory usually evolves from
detailed analyses of specific systems, a deep understanding
of a specific system may fail to materialize until a general theory has been developed. That, at least, was our ex-
perience: we did not fully understand the *-calculus until
we had achieved a general theory of the non-Newtonian
calculi. Nevertheless, Goethe was right: "...he who vivid-
ly grasps the particular will...also grasp the universal..."[2]

At this date we can only speculate as to the fruit-
fulness of the non-Newtonian calculi. Perhaps they can be
used to define new scientific concepts, to yield simpler
scientific laws, to solve heretofore unsolved problems.
The following remarks by Howard Eves are pertinent.

> "There seem to be two kinds of discovery. In
> one kind, the goal is given first and then
> the mind goes from the goal to the means,
> that is, from the question to the solution.
> In the other kind, the mind goes from the

means to the goal, that is, the mind first
discovers a fact and then seeks a use for
it. In mathematics, and elsewhere, most
significant discoveries are of the second
kind. As Hadamard has put it, 'Practical
application is found by not looking for it,
and one can say that the whole progress of
civilization rests on that principle.' An
outstanding example in mathematics is the
exhaustive study of the conics by the Greeks,
and then, some two thousand years later,
Kepler's stunning application of the Greek
findings to the movement of the planets in
the solar system. The physicist and artist
Duhem once compared Hadamard to a landscape
painter who in his studio creates a landscape painting and then leaves the studio to
find in nature some landscape fitting his
painting."[3]

In his beautiful Thematic Origins of Scientific
Thought, Gerald Holton recalls a tale told by the physicist
C. N. Yang to describe "the feelings of a physicist when he
consults a mathematician [or some of them at least]":

A man carried a large bundle of dirty clothes
and searched for a laundry without success
for a long time. He was greatly relieved
when he finally found a shop displaying a
sign 'Laundry done here'.... He went in and
dumped the bundle on the counter. The man
behind the counter said:

"What's this?"

"I want to have these laundered."

"We don't do laundry here."

"But you have a sign in the window
advertising that you do laundry."

"Oh! That! We only make signs."[4]

Well, we are not disdainful of the physicists' problems, and we shall be pleased to respond to any scientist
who seeks our counsel.

In the three centuries since the creation of the
classical calculus, the greatest mathematician was Gauss,
with whose wise words I conclude this book.

"In general the position as regards all such new calculi is this — That one cannot accomplish by them anything that could not be accomplished without them. However, the advantage is, that, provided such a calculus corresponds to the inmost nature of frequent needs, anyone who masters it thoroughly is able — without the unconscious inspiration of genius which no one can command — to solve the respective problems, yea, to solve them mechanically in complicated cases in which, without such aid, even genius becomes powerless. Such is the case with the invention of general algebra, with the differential calculus, and in a more limited region with Lagrange's calculus of variations, with my calculus of congruences, and with Möbius's calculus. Such conceptions unite, as it were, into an organic whole countless problems which otherwise would remain isolated and require for their separate solution more or less application of inventive genius."[5]

N O T E S

1. This perceptive remark was made by R. Duncan Luce and Patrick Suppes in their article "Mathematics," which appears in Volume 10 of the International Encyclopedia of the Social Sciences, to which a reference was made in Note 1 of Section 2.1.

2. Goethe expressed this thought on several occasions. Possessed by a distinct anti-mathematical bias, Goethe nevertheless made many interesting remarks about mathematics, one of which was considered profound by Oswald Spengler:
"The mathematician is only complete insofar as he feels...the beauty of the true."

3. This appears on pages 167, 168 of Howard Eves' entertaining book, Mathematical Circles Squared (Prindle, Weber, and Schmidt, 1972).

4. Holton's book is subtitled Kepler to Einstein and was published in 1973 by Harvard University Press.

5. Gauss' remark appears in Robert E. Moritz's book, On Mathematics and Mathematicians (Dover reprint, 1958), pp. 197-8. The Möbius calculus referred to by Gauss is the so-called barycentric calculus, which is geometry, not calculus in the usual sense.

LIST OF SYMBOLS

I N D E X

BIBLIOGRAPHY

Grossman, M. and Katz, R. *Non-Newtonian Calculus*. Rockport, MA: Mathco, 1972.
Included in this book, which was the first publication on non-Newtonian calculus, are discussions of nine specific non-Newtonian calculi, the general theory of non-Newtonian calculus, and heuristic guides for the application thereof.

Meginniss, J. R. "Non-Newtonian Calculus Applied to Probability, Utility, and Bayesian Analysis." *Proceedings of the American Statistical Association*: Business and Economics Statistics Section (1980), pp. 405-410.
This paper presents a new theory of probability suitable for the analysis of human behavior and decision making. The theory is based on the idea that subjective probability is governed by the laws of a non-Newtonian calculus and one of its corresponding arithmetics.

Grossman, J., Grossman, M., and Katz, R. *The First Systems of Weighted Differential and Integral Calculus*. Rockport, MA: Archimedes Foundation, 1980.
This monograph reveals how weighted averages, Stieltjes integrals, and derivatives of one function with respect to another can be linked to form systems of calculus, which are called weighted calculi because in each such system a weight function plays a central role.

Grossman, J. *Meta-Calculus: Differential and Integral*. Rockport, MA: Archimedes Foundation, 1981.
This monograph contains a development of systems of calculus, called meta-calculi, that transcend the classical calculus, for example in the following manner. In each meta-calculus the gradient, or average rate of change, of a function f on an interval [r,s] depends on all the points (x,f(x)) for which $r \leq x \leq s$, whereas the classical gradient $[f(s) - f(r)]/(s - r)$ depends only on the endpoints (r,f(r)) and (s,f(s)). The meta-calculi arose from the problem of measuring stock-price performance when taking all intermediate prices into account.

Grossman, M. *Bigeometric Calculus: A System with a Scale-Free Derivative*. Rockport, MA: Archimedes Foundation, 1983.
This book contains a detailed treatment of the bigeometric calculus, which has a derivative that is scale-free, i.e., invariant under all changes of scales (or units) in function arguments and values. Also included are heuristic guides for the application of that calculus, and various related matters such as the bigeometric method of least squares.

Grossman, J., Grossman, M., and Katz, R. *Averages: A New Approach*. Rockport, MA: Archimedes Foundation, 1983.
This monograph is primarily concerned with a comprehensive family of averages (unweighted and weighted) of functions that arose naturally in the development of non-Newtonian calculus and weighted non-Newtonian calculus. The monograph also contains discussions of some heuristic guides for the appropriate use of averages and an interesting family of means of two positive numbers.

www.ingramcontent.com/pod-product-compliance
Lightning Source LLC
Chambersburg PA
CBHW032013190326
41520CB00007B/454